Lecture Notes in Earth Sciences   87

Editors:
S. Bhattacharji, Brooklyn
G. M. Friedman, Brooklyn and Troy
H. J. Neugebauer, Bonn
A. Seilacher, Tuebingen and Yale

W0071818

Springer-Verlag Berlin Heidelberg GmbH

Michael Tiefelsdorf

Modelling
Spatial Processes

The Identification and Analysis
of Spatial Relationships
in Regression Residuals
by Means of Moran's I

With 32 Figures and 8 Tables

 Springer

Author

Prof. Dr. Michael Tiefelsdorf
The University of Manitoba, Department of Geography
211 Isbister Building, Winnipeg, Manitoba, Canada R3T 2N2

"For all Lecture Notes in Earth Sciences published till now please see final pages of the book"

Cataloging-in-Publication data applied for

Die Deutsche Bibliothek - CIP-Einheitsaufnahme

Tiefelsdorf, Michael:
Modelling spatial processes : the identification and analysis of spatial relationships in regression residuals by means of Moran's I ; with 8 tables / Michael Tiefelsdorf. - Berlin ; Heidelberg ; New York ; Barcelona ; Hong Kong ; London ; Milan ; Paris ; Singapore ; Tokyo : Springer, 2000
  (Lecture notes in earth sciences ; 87)

ISBN 978-3-540-66208-2   ISBN 978-3-540-48677-0 (eBook)
DOI 10.1007/978-3-540-48677-0

© Springer-Verlag Berlin Heidelberg 2000
Originally published by Springer-Verlag Berlin Heidelberg New York in 2000

Typesetting: Camera ready by author
SPIN: 10679990   32/3142-543210 - Printed on acid-free paper

This book is dedicated to the memory of my mother Ursula Tiefelsdorf

Preface

In recent years the analysis of spatial data has been proliferating. This is due to the recognition of the importance of locations and their spatial relations as a key to understanding many social, economic and environmental phenomena as well as their underlying processes. Also progresses in computing technology, in particular the microcomputer revolution, and developments of appropriate software, especially the advent of powerful, yet user-friendly, Geographic Information Systems, are supporting this trend. Furthermore, the broad availability of geo-referenced data, such as remotely sensed images as well as spatially disaggregated statistics by government and public organizations, mandates spatial data analysis. This monograph is a statistical contribution to this development which will enhance our understanding of spatial pattern and their underlying processes.

This monograph, which originated from my Ph.D.-research at Wilfrid Laurier University (Waterloo, Ontario, Canada), is about their analysis. It discusses the use of the exact distribution of the spatial autocorrelation test-statistic Moran's I. This work provides the statistical foundation of Moran's I for theoretically guided research on spatial processes. Moreover, it demonstrates the use of the proposed methodology by some practical applications and provides directions into which the methodology and empirical applications can be extended.

Some of these methodological directions are the interpretation and specification of local Moran's I_i as measure of hot spots, spatial clusters and spatial heterogeneities. In order to accomplish this, we need to understand the correlation among local Moran's I_is at several reference locations under the influence of a global spatial process. Another problematic applied area is the generalization of an observed global and local spatial pattern to identify its underlying spatial process or specific spatial singularities. For identification and generalization, we require several comparable realizations of an underlying spatial process. Some underlying statistical assumptions are difficult to test. Also on the technical front, improvements of our algorithms and their implementations, as well as sensible statistical approximations, are needed. – Research on spatial statistics and spatial autocorrelation is rapidly moving forward. Consequently, I would like to invite the reader to join me in the effort to stretching our boundaries of knowledge of this fascinating and relevant field of research.

Acknowledgments

Without the help and support of many individuals and organisations, it would have been impossible to complete this research on spatial autocorrelation and the distribution of the spatial autocorrelation coefficient Moran's I. First of all in Canada, Barry Boots of Wilfrid Laurier University deserves heartfelt thanks. Without our scholarly discussions within a stimulating atmosphere of academic freedom, this research would have never developed to this point. An initial scholarship by the International Council for Canadian Studies as well as financial support by Wilfrid Laurier University enabled me to begin working on this project. I owe both organizations many thanks. Furthermore, a grant by my university, The University of Manitoba, to produce the camera-ready manuscript is highly appreciated. I am very grateful for the multitude of support of this research in Berlin, Germany, especially from Gerhard Braun of the Free University of Berlin and from Dieter Schön of the Robert-Koch-Institute. Furthermore, I would like to thank all my friends on both sides of the Atlantic Ocean and, in particular, my wife Jane Duffy Tiefelsdorf, for their patience and encouragement during the assembly of this book.

I gratefully acknowledge the following publishers for their permissions to reproduce various figures: The Regional Science Association International permitted the replication of Figs. 10.4, 10.5, 11.1, 11.2, 11.3 and 11.8 (*Papers in Regional Science*, Vol. 77, No. 2, April 1998). Pion Limited, London, granted permission to reproduce Figs. 7.2, 8.1 and 8.2 (*Environment and Planning A*, Vol. 27, pages 985–999, 1995). Ohio State University Press permitted the use of Fig. 11.6 (*Geographical Analysis*, Vol. 29, No. 3, July 1997). As usual, the copyright for the figures listed above remains property of these publishers.

Table of Contents

List of Tables

List of Figures

Chapter 1

Introduction

To set the *theoretical* and *methodological* frame of this monograph, neither spatial structure nor spatial autocorrelation can exist independently from each other. Without an underlying spatial structure spatially autocorrelated phenomena cannot diffuse and without the presence of significant spatial autocorrelation spatial structure becomes meaningless from an empirical point of view. These two interwoven concepts together determine the ingredients of any spatial and/or temporal stochastic process. The presented perspective on spatial structure, spatial autocorrelation and a spatial process is in line with Haining's (1990, p 24) view on spatial processes who states that "spatial structure arises from the operation of processes in which spatial relationships enter explicitly into the way the process behaves". While the specification of spatial structures has to be theoretically guided by the hypothesized spatial process, the spatial process itself finds its legitimacy in the presence of a significant spatial autocorrelation level.

This theoretically guided approach toward spatial structure and spatial autocorrelation is not novel at all. Nevertheless, it does not follow the mainstream treatment of spatial autocorrelation which either explicitly or implicitly regards it as an expression of many different forms of theoretical or operational misspecifications. See for instance Griffith (1992), who lists a battery of reasons for, and expressions of, spatial autocorrelation related to misspecification with only once coming close to the process-based *modus operandi* of it by attaching it to "spatial spill-over effects". Surprisingly, without claiming absoluteness, it is epidemiologists and medical geographers who either deliberately or instinctively embraced the notion of a spatial process in their studies on geo-referenced health information by looking at spatial diffusion of contagious diseases. We owe this process-based perspective in medical geography to the pioneering work of Haggett (1976) and Cliff and Ord (1981). To follow this geographical tradition an epidemiological study on the spatial distribution of cancer incidences and their underlying processes will set the *applied* and *empirical* frame of this monograph.

It is this process-tied perspective on spatial structure and spatial autocorrelation which this book will adopt. Spatial structure is regarded here as a linking substance and functional connection between interrelated spatial objects on which the spatial process evolves with a given strength. To structure and condense the empirical diversity and to make it accessible to scientific investigations, epistemologically the observed spatial objects are regarded as appropriate aggregates of elementary spatial entities and spatial relations are reflected by meaningful abstractions of a functional spatial order between the spatial objects. An observed map pattern or spatial distribution, respectively, is supposed to be a random manifestation of many possible representations of the hypothetical spatial process. This necessary simplification of the empirical diversity is bought at the price of methodological problems. Some of these will be addressed in this work.

The intention of this work is not to catalogue all possible spatial processes, but to use some specific hypothetical spatial processes in order to demonstrate the feasibility and flexibility of a newly introduced *closed statistical theory*. Its purpose lies in the identification of an underlying spatial process and the investigation of local non-stationarities based on residuals from a regression model. This theory has its roots in the exact distribution of ratios of quadratic forms in normal variables. The widely used spatial autocorrelation coefficient Moran's *I* is one member of these ratios. In distinction to available test statistics for spatial autocorrelation, which are only asymptotically valid and require spatial independence, the presented method is exact for finite samples even under a significant hypothetical spatial process. The method can not only be used in hypothesis testing, it is also helpful in exploratory spatial data analysis which can be implemented in Geographical Information Systems. In particular, under a significant spatial process or specific forms of the spatial structure the asymptotic distributions of the test statistics for spatial autocorrelation are not available. Consequently, so far one needs to retreat to simulation experiments which require substantial knowledge from the analyst to design the experiment appropriately in the quest of obtaining statistically reliable results.

1.1   The Origins of the Exact Approach

This section carves out the main roots of the exact tests on spatial dependence dating back to vibrant developments in econometrics in the Fifties, Sixties and Seventies. By far, this discussion is not intended to be exhaustive and I would like to apologize to all those scholars whose contributions have not been explicitly mentioned here. Without their engagements the present stream of knowledge only would resemble a rivulet. Furthermore, the people fostering progress and dissemination of computing technology have to be credited. Coming from a mainframe era and having started with punch cards the microcomputer revolution came as an act of liberation – an act which is still in its full swing.

Notably the theoretical work by Durbin and Watson (1950, 1951, and 1971) on a test statistic for serial autocorrelation defined as ratio of quadratic forms in regression residuals broke ground to tackle a problem troubling statisticians for over many decades. King (1992) gives an excellent account of the historical background on Durbin and Watson's work, the constraints under which it was developed, and the significance of Durbin and Watson's results. In the Fifties computing was still performed by mechanical machines, therefore, Durbin and Watson had to employ simplifying assumptions. These led to bounding significance points of their statistic which do not depend on the design matrix. The bounding significance points were approximated by the beta-distribution based on the first two moments of their statistic and its feasible range. In a separate development Imhof (1961) derived a practical approach to calculate by numerical integration the exact distribution of a sum of central and non-central χ^2-distributed variables. Abrahamse and Koerts (1969) and Koerts and Abrahamse (1968 and 1969) synthesized the results of Durbin and Watson and Imhof and developed the econometric theory further. Their pivotal contribution was the algebraic reduction of the distribution of a ratio of quadratic forms into a distribution of a simple quadratic form, i.e., a sum of χ^2-distributed variables. This allowed them to calculate the exact distribution of Durbin and Watson's statistic under the assumption of serial independence and conditional on a significant temporal process.

The work of Durbin and Watson, Imhof, and Abrahamse and Koerts laid the foundation for advancing econometric research on serial autocorrelation in the linear model. Notably King's contributions and bibliographic reviews have to be mentioned here. While his work focused mainly on the serial aspects of economic datasets, he was well aware of the cross-sectional aspects, too (King, 1981, Evans and King, 1985). The acknowledgment of spatial aspects in socio-economic datasets is still quite uncommon in econometric applications. His survey article on testing for autocorrelation in linear regression models with 245 references gives a peerless account on the state of art until the mid Eighties (King, 1987). While King covered the econometric and statistical issues, Farebrother's ongoing contribution is the critical dissemination of numerical solutions for calculating the exact distribution of quadratic forms.

Surprisingly, little recognized in the econometric literature but nonetheless important is the work by Sawa (1972 and 1978) on the moments of ratios of quadratic forms in normal variables under the presence of a significant temporal process. The devised formulas for the first four moments under the alternative hypothesis are also applicable to numerical integration and easy to implement. Somehow this negligence of the conditional distribution under the alternative hypothesis in the econometric practice is comprehensible. Due to the uni-lateral and uni-directional character of time-series the distribution of the test statistics under a alternative hypothesis is mainly used for power investigations. However, in the spatial situation the spatial structure and its specification under the zero hypothesis (spatial process under investigation) and an alternative hypothesis (spatial reference process) can differ which gives a much more flexible framework for investigations.

Without those econometric and statistical developments, the present work on exact small sample spatial autocorrelation analysis would not be possible. The top block of Figure 1.1 denotes this foundation of present developments in spatial analysis. The statistical results of Abrahamse and Koerts as well as of Durbin and Watson emanated into the work of Cliff and Ord on spatial autocorrelation. One contribution of Cliff and Ord as well as Haggett is their remarkable inauguration of spatial relationship and tessellation models to spatial and regional science for the evaluation of spatial processes. This introduction of spatial relationship and tessellation models paved the way for the present practice in spatial analysis and spatial econometrics and is a valuable source for ongoing research. Tessellation models have to be distinguished from distance based models which fall into the domain of geo-statistics.

The initiative of Cliff and Ord as well as Haggett carried over to Haining and Anselin besides many other spatial and social scientists in the late Seventies and early Eighties. While Haining's work is more in line with the tradition in spatial statistics, Anselin's viewpoint draws heavily from the developments in econometrics. Their activities were embedded in the asymptotic maximum likelihood framework based on a large sample theory and in simulation studies. Also Bartels and Hordijk (1977) followed the simulation approach to understand the behavior of spatial test statistics, however, they were the first to mention in spatial statistics the possibility of approaching the distribution of test statistics on spatial association under the zero and alternative hypotheses by exact methods.

With the increasing availability of desktop computing power and numerical software it was a consequent step further in the middle of the Nineties to re-discover the exact methods to investigate the properties of spatial autocorrelation tests. All these methods require numerical integration and the calculation of eigenvalues for large matrices reflecting the relationships between the spatial objects. The first application of these methods to calculate the exact distribution of Moran's I was given by Tiefelsdorf and Boots in 1994 at the American Association of Geographers annual meeting in San Francisco. Almost simultaneously Hepple started re-considering the exact distribution of Moran's I based on previous thoughts in his unpublished Ph.D.-thesis in 1974. In a series of papers Tiefelsdorf and Boots developed the exact method further. The distribution of Moran's I conditional on a significant spatial process and the conditional moments were applied under several circumstances by Tiefelsdorf (1998). For instance, the exact distribution of local Moran's I_i under the zero hypothesis and conditional on a significant global spatial process was given in 1995. The additivity property of the local Moran's I_i's (Anselin, 1995) to match global Moran's I allowed Tiefelsdorf, Griffith and Boots (1999) to investigate the impact of different coding schemes on the significance of the global statistics. As a consequence of the topology induced heterogeneity in the local variances they devised a variance stabilizing coding scheme.

One can expect that within the next few years more applications using the exact method or modifications of it will become available and thus the box "Exact Approach by Numerical Integration" in Figure 1.1 will fill up. Nowadays, several hundred simultaneously linked spatial objects can be handled without problems on well equipped personal computers. What high end computers are nowadays will be tomorrow's scrap, therefore we can expect for the future even more sophisticated models to become standard repertoires in empirical analysis. The advantage of the exact method is that it can be implemented in statistical packages as a black box whereas simulation studies always require substantial efforts to be setup. The potential for applications of the conditional distribution of spatial autocorrelation statistics is by far broader than in time-series analysis because the spatial link matrix under the zero hypothesis and under the alternative hypothesis can differ. As simulation studies and the maximum likeli-

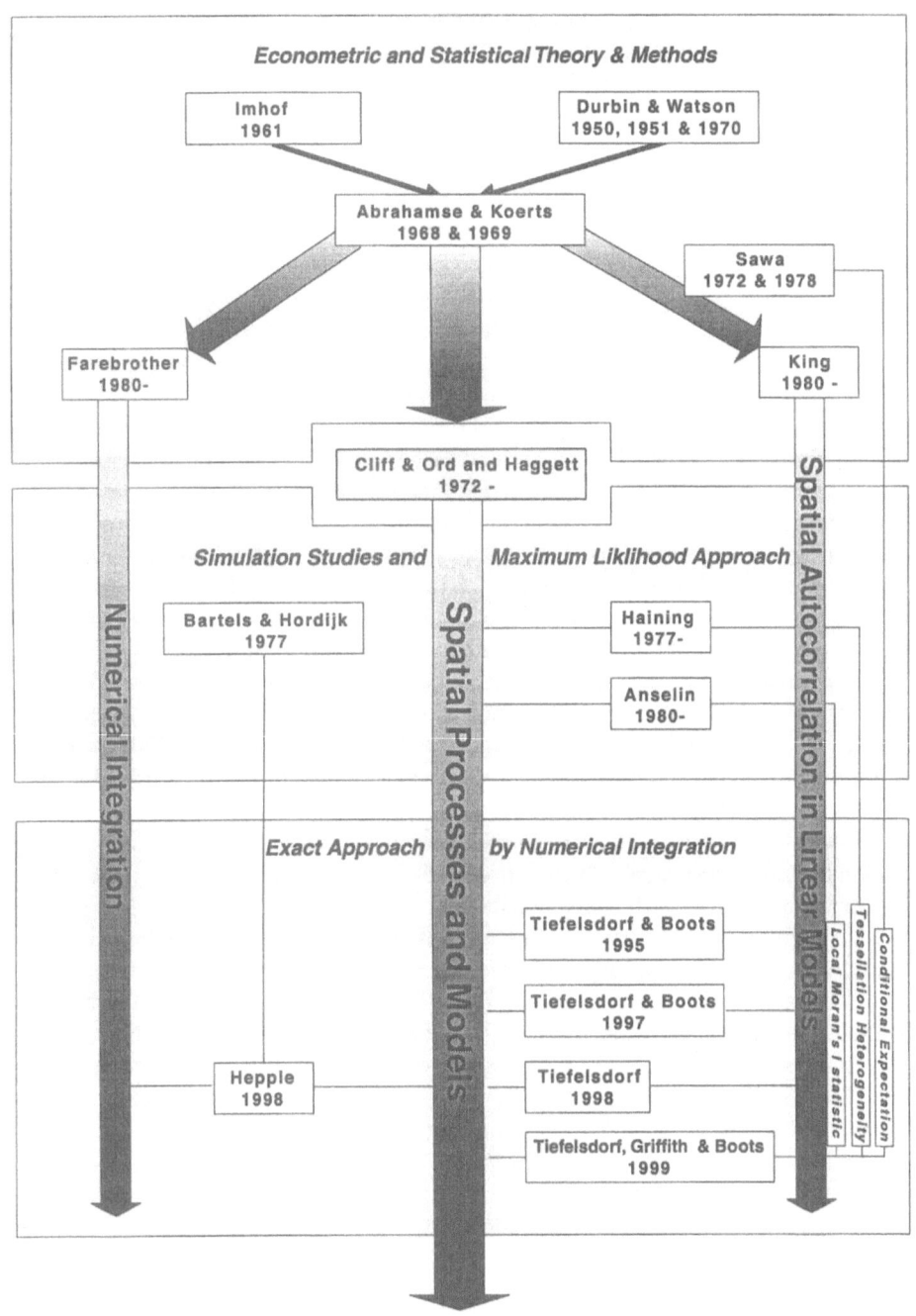

Fig. 1.1. Genealogy leading to exact statistical tests in spatial analysis

hood approach sustain the progress in handling and modeling spatial processes, so it can be anticipated that also the exact approach will nurture the body of knowledge on spatial processes.

1.2   Organization of the Monograph

The monograph is organized in three main parts. In Part I the basic concept of a stochastic spatial process in regression disturbances will be introduced. Therefore, first some basic properties of the ordinary regression model are summarized and the violation from the assumption of independence in the disturbances is discussed. Reasons are given why a misspecified regression model can lead to spatial autocorrelation. Then the concept of spatial structure is introduced. It is based on the Cartesian product of the n spatial objects as functional relationships. These relationships are specified either as simple Euclidean distances or as a relational spatial graph. While distance relations are of prime relevance for the investigation of natural phenomena, human activities rarely follow a simple distance decay principle. Contiguity graphs and their associated contiguity matrices are the binary analogue to distance matrices. Other forms presented are higher order link matrices expressing adjacencies of pairs of spatial objects p links apart on the contiguity graph. In general terms also functional relationships in relative space are discussed, such as spatial hierarchies, a prominent planar spatial structure in human and economic geography, or non-planar spatial networks. Weighted networks play an important role in the investigation of spatial relationships such as interregional interaction models or of social relationships between individuals and/or groups (see Doreian, 1974, and Doreian et al., 1984).

Besides the specification of the spatial relation matrix also the coding of the single links is of relevance. The two prominent coding schemes are the global standardization of the spatial link matrix and the row-sum standardization, where each row in the link matrix is adjusted individually. Due to the differential linkage degree of the spatial objects these coding schemes have a marked impact on the results of spatial analysis. A novel coding scheme coping with this topology induced heterogeneity is introduced (see Tiefelsdorf, Griffith and Boots, 1999). Another manipulation of the global spatial relation matrix is its decomposition into the local components. These local link matrices give rise to local autocorrelation statistics (see Tiefelsdorf and Boots, 1997). These recently introduced statistics are used as an exploratory tool to detect spatial heterogeneity in spatial processes. It is important to know the properties of these local autocorrelation statistics and their relation to the global autocorrelation statistic.

Next in Part I the concept of spatial autocorrelation in Gaussian spatial processes is introduced. The spatial structure and the spatial autocorrelation coefficient are set into relationship to each other. Basic assumptions in modeling general spatial processes are discussed. Furthermore, reasons will be given why we observe spatial autocorrelation in our data under investigation. Due to the lack of an underlying theory mostly the plain spatial adjacency structure is used which indicates in misspecified models spatial autocorrelation. Then the different

Gaussian spatial processes and their siblings are discussed together with their properties.

Part II takes most of the space of this book. It mainly deals with technical issues leading to a closed theory on the exact distribution of Moran's I. In general terms the exact distribution of quadratic forms in normal variables is derived and a method for its computation is given. It will be shown then how to apply this method to calculate the exact distribution and the significance points of Moran's I under either the assumption of spatial independence or conditional on a significant Gaussian spatial process. In addition, I will give under both conditions the feasible range of Moran's I, its statistical moments as well as the correlation between Moran's Is defined on different specifications of their underlying spatial structures. Also a comparison between approximation methods, such as the normal approximation, and the exact distribution of Moran's I is given.

Furthermore, Part II addresses the effects of the negligence of the design matrix on tests for spatial autocorrelation and introduces bounding distributions within which the exact significance point must lie. These bounding distributions also allow one to assess asymptotic maximum likelihood tests which disregard the effects of the design matrix and to test residuals from maximum likelihood estimation procedures for spatial autocorrelation in other than the reference process.

Part III takes up ideas from the previous parts and deals with the application of the concepts and methods introduced there. First a epidemiological data set on male bladder cancer is described briefly. It stems from the cancer registry of the former German Democratic Republic, which in its completeness and focus on age standardized incidences instead of mortalities is unique world-wide. A process based on spatial contiguity will be investigated for heterogeneity in the form of local clusters or local outliers in the cancer incidences by means of local Moran's I_i. This is followed by a non-nested identification procedure of whether the spatial distribution of the cancer incidences is more in-line with a process based on spatial contiguity or with a process following a hierarchical spatial migration pattern.

An outlook on the usefulness and applicability of the exact approach of testing for spatial processes as well as some comments on disease mapping closes the monograph. My hope is that by stressing this process-based perspective of spatial autocorrelation analysis, fosters its dissemination into the body of commonly used geographical research tools.

The algorithms used throughout this research and their coding as well as the links to Geographic Information Systems are still in a prototype state thus a presentation of specific computer programs is not given. All numerical calculations were performed by using GAUSS 3.2 (Aptech Systems, 1996) because of its high performance and efficiency for vectorised data structures. For basic statistical data analysis and non-spatial data manipulations SPSS 6.1.3 (SPSS Inc., 1995) was used. Some analytical results were verified with the support of MAPLE 4.0, as implemented in the SCIENTIFIC WORKPLACE 2.5 (ITC Software Research,

1996). All spatial analyses, Geographic Information System operations, digitizing and mapping tasks were done with MAPTITUDE 4.0 (Caliper Inc., 1997). Some user and data interface macros were written in MAPTITUDE'S DEVELOPMENT TOOLKIT. Spatial attribute data were exchanged between the programs in the DBase-format. Despite the prototype character of the programs the exact method to derive the distribution of Moran's I can already be used on personal computers. I envisage that in a few years with increasing personal computing power and open Geographic Information Systems exact methods will be become part of the standard scientific repertoire.

Part I

Underlying Concepts of Spatial Processes

This Part will deal with two essential groups of information used to *grasp* spatially distributed phenomena, to *formulate* and *test* hypotheses on their generating mechanism, and finally to *model* this mechanism which allows one to conduct spatial control and predictions. These two groups of information are the *deterministic* observable factors and the invisible *stochastic* components setting up an underlying spatial process. The discussion given here is restricted to simple regression models and plain Gaussian spatial processes. In the regression model the hypothetical *explanatory variables* and the underlying hypothetical *spatial structure* are assumed to be deterministic whereas the *endogenous variable* follows a random distribution via the disturbances. The distribution of the disturbances is generated by the stochastic spatial process utilizing the underlying spatial structure.

This logical decomposition into a *random* and a *deterministic* component sets the stage for the following chapters. First the ordinary least squares regression (OLS) model is introduced and its properties under the influence of a spatial process or missing spatially autocorrelated variables are investigated. Then I move to the fundamental building block of a spatial process, that is, the spatial structure characterizing the hypothetical relations between the spatial objects. Furthermore, some notable specifications of the spatial structures and different coding schemes are presented in this context. With the spatial structure at hand, spatial processes are defined as a mechanism which are linking the so far independent white noise disturbances together spatially. While the perspective on spatial structures taken here is more narrow and disregards plain *geometrical arguments*, it is sufficiently abstract to set up a closed and general theory with *empirical content*. Finally, statistical test principles to determine the presence of an underlying spatial process are discussed.

It is evident that this syllabus focuses on the underlying *spatial structure* and *distributional issues* to understand the underlying *stochastic mechanism* which generates an observed sample attached to geo-referenced spatial objects. Consequently, simultaneous modelling and estimating techniques towards spatial regression models by means of the maximum likelihood approach will be neglected.

Chapter 2

The Regression Model

Epistemologically the quest of empirical sciences is to reveal hidden knowledge on empirical phenomena. In an initial stage this knowledge may be hypothetical and derived from underlying theoretical considerations. Once this knowledge has been uncovered and substantiated to reflect reliably our social and physical environment, it will becomes part of society's erudition. From a researcher's point of view the revealed systematic information loses its scientific bearing. What counts for a researcher is the so far uncovered information remaining hidden in the data. This hidden information initially appears to be disordered to the researcher. Does it tell us more about the phenomena under investigation? It is here, where progress lies.

This differentiation between an explained component and the unsystematic information is intrinsic to the regression model. There is the systematic component consisting of exogenous information tied together by a set of parameters expressing the expectation of an observation and there is its unexplained regression residual. If the regression residuals are totally at random there is no more information to excavate from the data. However, if they exhibit, up to stochastic variations, an underlying structure, this structure can tell us something about the underlying mechanism which generated the data.

At first the classical OLS regression model will be reviewed and the properties of its residuals are summarized. Then the classical regression model will be evaluated under the perspective that the residuals exhibit some sort of structure expressed by a covariance matrix of the disturbances. This chapter will be closed by a discussion of a mis-specified regression model where some relevant exogenous variables have not been considered. The following discussion rests substantially on the econometric exposition by Fomby et al. (1984).

For most parts, this chapter does not rely on any specific distribution assumption of the disturbances and simply assumes a specific form of the first and second moments of the parameters and residuals. However, if statistical tests have to be performed an exact distribution model is required. It is standard practice to assume then that the disturbances follow a multivariate normal distribution. This is the distribution model applied throughout this book.

2.1 A Review of Ordinary Least Squares

The classical linear regression model gives a description on population properties of a random variable Y. It is assumed that the n spatially distributed observations $y_1, y_2, \ldots, y_n$ on Y constitute a random sample where the index of the observations denotes also their location. Their means are conditional to a linear combination of k non-stochastic variables $\mathbf{x}_1, \mathbf{x}_2, \ldots, \mathbf{x}_k$ where the first variable is usually assumed to be the $n \times 1$ constant unity vector $\mathbf{x}_1 \equiv \mathbf{1} = (1, 1, \ldots, 1)^T$. The set of the k regressors is also called the *design matrix* and denoted by $\mathbf{X} \equiv (\mathbf{x}_1, \mathbf{x}_2, \ldots, \mathbf{x}_k)$. More precisely,

Definition 1 (The Classical Linear Regression Model). *The classical regression model is specified by*

$$\mathbf{y} = \mathbf{X} \cdot \boldsymbol{\beta} + \varepsilon \tag{2.1}$$

where $\mathbf{y} = (y_1, y_2, \ldots, y_n)^T$, $\varepsilon = (\varepsilon_1, \varepsilon_2, \ldots, \varepsilon_n)^T$, $\boldsymbol{\beta} = (\beta_1, \beta_2, \ldots, \beta_k)^T$,

$$\mathbf{X} = \begin{pmatrix} x_{11} & x_{12} & \cdots & x_{1k} \\ x_{21} & x_{22} & \cdots & x_{2k} \\ \vdots & \vdots & \ddots & \vdots \\ x_{n1} & x_{n2} & \cdots & x_{nk} \end{pmatrix} \tag{2.2}$$

and the following assumptions are satisfied

(i) $\mathbf{X}$ *is a non-stochastic matrix of rank* $k \leq n$ *and, as the sample size becomes infinitely large,* $\lim_{n \to \infty}(\frac{1}{n}\mathbf{X}^T\mathbf{X}) = \mathbf{Q}$, *where* $\mathbf{Q}$ *is a finite and non-singular matrix.*

(ii) *The disturbance vector* ε *consists of unobservable random errors which satisfy the properties* $\mathrm{E}[\varepsilon] = \mathbf{0}$ *and* $\mathrm{E}[\varepsilon \cdot \varepsilon^T] = \sigma^2 \cdot \mathbf{I}$ *where* $\mathrm{E}[\cdot]$ *denotes the mathematical expectation and* $\mathbf{I}$ *is the identity matrix.*

Assumption (i) limits the behavior of the regressors $\mathbf{X}$. The regressors can not increase as fast as the number of observations n. Also none of the regressors may become zero or collinear in the limit. Furthermore, assumption (i) excludes collinearity among the regressors because of the full column rank assumption on $\mathbf{X}$, where the number of observations n is larger than the number of regressors k. Note that assumption (ii) does not require a specific form of the distribution for the disturbance vector ε. It simply requires that their expectations are zero and that the single disturbances are mutually independent. Foremost, the conditional mean of the endogenous variable $\mathbf{y}$ subject to the exogenous variables $\mathbf{X}$ is given by $\mathrm{E}[\mathbf{y} \mid \mathbf{X}] = \mathbf{X} \cdot \boldsymbol{\beta}$ since by assumption $\mathrm{E}[\varepsilon] = \mathbf{0}$ and the independence of disturbances ε from the non-stochastic design matrix $\mathbf{X}$. The conditional expectation $\mathrm{E}[\mathbf{y} \mid \mathbf{X}]$ reflects our systematic knowledge about a phenomenon $\mathbf{y}$ under investigation whereas the disturbances ε carry the unknown components which are out of reach of the researcher.

Because the endogenous variable $\mathbf{y}$ is additively linked to the disturbances ε it is also a random function. Thus it can be observed only up to the inherently so far uncontrollable random shocks ε. Likewise the regression parameters $\boldsymbol{\beta}$, which tie the conditional expectations $\mathrm{E}[\mathbf{y} \mid \mathbf{X}]$ to the observed regression matrix, are not directly available. However, for the estimates of the regression parameters $\boldsymbol{\beta}$ by the *ordinary least squares estimator* (OLS)

$$\hat{\boldsymbol{\beta}} = (\mathbf{X}^T\mathbf{X})^{-1}\mathbf{X}^T\mathbf{y} \tag{2.3}$$

the Gauss-Markov theorems holds some strong results in stock:

Theorem 1 (Gauss-Markov). *Assume the properties of the classical linear regression model hold. The ordinary least squares estimator* $\widehat{\beta}$ *is unbiased and efficient within the class of linear, unbiased estimators.*

Unbiasness and efficiency means:

Definition 2 (Unbiasness). *Let* β^* *be an estimator of* β; *then* β^* *is unbiased if* $\mathrm{E}[\beta^*] = \beta$.

Definition 3 (Efficiency). *An unbiased estimator is efficient with respect to a class of estimators if it has a variance which is not greater that the variance of any other estimator in that class.*

Consequently, the ordinary least squares estimator $\widehat{\beta}$ is "on average over all potential samples" identical to the underlying regression parameter β and its variance is smaller than that of any other linear, unbiased estimator $\beta^* = \mathbf{H} \cdot \mathbf{y}$ where $\mathbf{H}$ is defined by $\mathbf{H} \equiv (\mathbf{X}^T\mathbf{X})^{-1}\mathbf{X}^T+\mathbf{C}$ and $\mathbf{C}$ a matrix of arbitrary constants. For details and a proof see Fomby et al. (1988, pp 10–11).

However, because the covariance of the ordinary least squares estimator $\widehat{\beta}$ is needed in the next section, it shall be given here. Due to its unbiasness, the covariance matrix of the OLS estimator $\widehat{\beta}$ is calculated under the independence assumption by

$$\mathrm{E}\big[\big(\widehat{\beta} - \mathrm{E}(\beta)\big) \cdot \big(\widehat{\beta} - \mathrm{E}(\beta)\big)^T \mid \sigma^2\mathbf{I}\big]$$

$$= \mathrm{E}[(\widehat{\beta} - \beta) \cdot (\widehat{\beta} - \beta)^T] \tag{2.4}$$

$$= \mathrm{E}\big[\big((\mathbf{X}^T\mathbf{X})^{-1}\mathbf{X}^T\mathbf{y} - \beta\big) \cdot \big((\mathbf{X}^T\mathbf{X})^{-1}\mathbf{X}^T\mathbf{y} - \beta\big)^T\big] \tag{2.5}$$

$$= \mathrm{E}\big[\big((\mathbf{X}^T\mathbf{X})^{-1}\mathbf{X}^T(\mathbf{X} \cdot \beta + \varepsilon) - \beta\big) \cdot \tag{2.6}$$

$$\big((\mathbf{X}^T\mathbf{X})^{-1}\mathbf{X}^T(\mathbf{X} \cdot \beta + \varepsilon - \beta\big)^T\big]$$

$$= \sigma^2 \cdot (\mathbf{X}^T\mathbf{X})^{-1} . \tag{2.7}$$

The Gauss-Markov theorem provides a small-sample result which is independent of the underlying distributional form of the disturbances ε.

The estimated regression residuals $\widehat{\varepsilon}$ measure the variation of the observed endogenous variable $\mathbf{y}$ around its estimated conditional expectation $\mathbf{X} \cdot \widehat{\beta}$, that is,

$$\widehat{\varepsilon} = \mathbf{y} - \mathbf{X} \cdot \widehat{\beta} . \tag{2.8}$$

Because $\dot{\mathrm{E}}[\widehat{\varepsilon}] = \mathrm{E}[\mathbf{X} \cdot \beta + \varepsilon - \mathbf{X} \cdot \widehat{\beta}] = \mathbf{X} \cdot \mathrm{E}[\beta - \widehat{\beta}] + \mathrm{E}[\varepsilon] = \mathbf{0}$ the regression residuals are unbiased, too, and have zero expectation. Using equations (2.3) and (2.8) the residuals can be expressed as

$$\widehat{\varepsilon} = \big(\mathbf{I} - \mathbf{X} \cdot (\mathbf{X}^T\mathbf{X})^{-1} \cdot \mathbf{X}^T\big) \cdot \mathbf{y} \tag{2.9}$$

where

$$\mathbf{M} \equiv \mathbf{I} - \mathbf{X} \cdot (\mathbf{X}^T\mathbf{X})^{-1} \cdot \mathbf{X}^T \tag{2.10}$$

is called the *projection matrix*.

As its name indicates, matrix $\mathbf{M}$ projects the endogenous variable $\mathbf{y}$ and the disturbances ε into the residual space which is orthogonal to exogenous variables $\mathbf{X}$, that is,

$$\hat{\varepsilon} = \mathbf{M} \cdot \mathbf{y} = \mathbf{M} \cdot \varepsilon , \tag{2.11}$$

$$\mathbf{0} = \mathbf{M} \cdot \mathbf{X} = \mathbf{M} \cdot \hat{\mathbf{y}} \text{ with } \hat{\mathbf{y}} = \mathbf{X} \cdot \hat{\beta} \text{ or} \tag{2.12}$$

$$\mathbf{0} = \mathbf{X}^T \cdot \hat{\varepsilon} . \tag{2.13}$$

Consequently, the regression residuals $\hat{\varepsilon}$ are orthogonal to the design matrix $\mathbf{X}$ which means that they are free from any systematic effect related to the design matrix. In particular, because the regression residuals are orthogonal to the unity vector $\mathbf{1}$ their mean is zero. Several other properties of the projection matrix are noteworthy. Its rank is

$$n - k = \text{rank}(\mathbf{M}) . \tag{2.14}$$

It is an idempotent, symmetric matrix, that is,

$$\mathbf{M} = \mathbf{M} \cdot \mathbf{M} , \tag{2.15}$$

and, consequently, it has $n - k$ unity eigenvalues and k zero eigenvalues. Due to the $n - k$ unity eigenvalues the trace of $\mathbf{M}$ is $n - k$.

If the design matrix $\mathbf{X}$ consists solely of the constant unity vector $\mathbf{1}$ then its special projection matrix $\mathbf{M}_{(1)}$ has the form

$$\mathbf{M}_{(1)} \equiv \begin{pmatrix} 1 - \frac{1}{n} & -\frac{1}{n} & \cdots & -\frac{1}{n} \\ -\frac{1}{n} & 1 - \frac{1}{n} & & -\frac{1}{n} \\ \vdots & & \ddots & \vdots \\ -\frac{1}{n} & -\frac{1}{n} & \cdots & 1 - \frac{1}{n} \end{pmatrix} . \tag{2.16}$$

It measures the most basic variation of the observation vector $\mathbf{y}$, that is, the variation around its common mean $\bar{y} = \frac{1}{n} \sum_{i=1}^{n} y_i$ by

$$\mathbf{y} - \bar{y} \cdot \mathbf{1} = \mathbf{M}_{(1)} \cdot \mathbf{y} . \tag{2.17}$$

Consequently, as long as a constant term $\mathbf{1}$ is included in the design matrix $\mathbf{X}$ the average regression residual will be zero.

While a fundamental assumption of ordinary least squares states that the disturbances are stochastically independent, that is, $\text{E}[\varepsilon \cdot \varepsilon^T] = \sigma^2 \cdot \mathbf{I}$, this no longer holds for the regression residuals $\hat{\varepsilon}$. By using property (2.11) and the idempotency of the projection matrix $\mathbf{M}$ their covariance becomes

$$\text{E}[\hat{\varepsilon} \cdot \hat{\varepsilon}^T] = \text{E}[\mathbf{M} \cdot \varepsilon \cdot \varepsilon^T \cdot \mathbf{M}] = \sigma^2 \cdot \mathbf{M} \tag{2.18}$$

Thus the regression residuals are no longer independent from each other and their dependence relies via the projection matrix $\mathbf{M}$ on the design matrix $\mathbf{X}$.

Consequently, any independence test must take this structural dependence between the observable regression residuals explicitly into account. This leads to complications for the calculation of the distribution of the test statistics and several courses out of this dilemma have been sought. Chapter 7.2.3 on the bounding distributions of Moran's I gives a detailed discussion on this subject. Note that asymptotically the impact of the design matrix $\mathbf{X}$ on the projection matrix $\mathbf{M}$ diminishes and the projection matrix $\mathbf{M}$ approaches the identity matrix $\mathbf{I}$, that is, $\lim_{n\to\infty} \mathbf{M} = \mathbf{I}$ (Upton and Fingleton, 1985, p 337).

2.2   Ordinary Least Squares with Correlated Disturbances

Before discussing the properties of an OLS estimator under the presence of correlated disturbances, the correct estimator for these non-spherical disturbances needs to be given. This allows me to compare the erroneously applied OLS estimators against the correct estimator which utilizes the information in the correlation of the disturbance. To mention it in advance, the OLS estimator will not lose two important properties, namely the unbiasness and the consistency. However, it will not be efficient anymore and predictions conducted with OLS are no longer best, linear and unbiased. Thus forecasts based on the wrong model are suboptimal.

In contrast to the presumption of OLS let us assume that the disturbances ε have a non-spherical covariance matrix of

$$E[\varepsilon \cdot \varepsilon^T] = \mathbf{\Omega} \tag{2.19}$$

where $\mathbf{\Omega}$ is a *known* symmetric, positive definite matrix. This general covariance matrix $\mathbf{\Omega}$ permits the possibility that the disturbances are correlated or have different variances. The covariance matrix $\mathbf{\Omega}$ can be decomposed into a product of two non-singular matrices by, for instance, a Cholesky decomposition, that is,

$$\mathbf{\Omega}^{-1} = \mathbf{P} \cdot \mathbf{P}^T . \tag{2.20}$$

Then Aitken's generalized least squares (GLS) theorem states

Theorem 2 (Aitken's Theorem). *The generalized least squares estimator*

$$\tilde{\beta} = (\mathbf{X}^T \cdot \mathbf{\Omega}^{-1} \cdot \mathbf{X})^{-1} \cdot \mathbf{X}^T \cdot \mathbf{\Omega}^{-1} \cdot \mathbf{y} \tag{2.21}$$

is efficient among the class of linear, unbiased estimators of β.

This estimator is equivalent to an OLS estimator in a transformed model $\mathbf{y}^* = \mathbf{X}^* \cdot \beta + \varepsilon^*$ with $\mathbf{y}^* = \mathbf{P}^T \cdot \mathbf{y}$, $\mathbf{X}^* = \mathbf{P}^T \cdot \mathbf{X}$ and $\varepsilon^* = \mathbf{P}^T \cdot \varepsilon$ where $E[\mathbf{P}^T \cdot \varepsilon \cdot \varepsilon^T \cdot \mathbf{P}] = \mathbf{P}^T \cdot \mathbf{\Omega} \cdot \mathbf{P} = \sigma^2 \cdot \mathbf{I}$. Thus GLS estimator has covariance $\text{Cov}[\tilde{\beta}] = (\mathbf{X}^T \cdot \mathbf{\Omega}^{-1} \cdot \mathbf{X})^{-1}$.

Under the assumption that $\lim_{n\to\infty}(\frac{\mathbf{X}^T \cdot \mathbf{\Omega} \cdot \mathbf{X}}{n}) = \mathbf{Q}_g$ is finite and non-singular the OLS estimator is still *unbiased* because

$$E[\hat{\beta}] = \beta + (\mathbf{X}^T \cdot \mathbf{X})^{-1} \cdot \mathbf{X}^T \cdot \underbrace{E[\varepsilon]}_{=0} = \beta , \tag{2.22}$$

and it is a *consistent* because

$$\text{plim}[\widehat{\boldsymbol{\beta}}] = \boldsymbol{\beta} + \lim_{n\to\infty} \left[\left(\frac{\mathbf{X}^T \cdot \mathbf{X}}{n} \right)^{-1} \right] \cdot \text{plim} \left[\frac{\mathbf{X}^T \cdot \varepsilon}{n} \right] . \tag{2.23}$$

The term $\frac{\mathbf{X}^T \cdot \varepsilon}{n}$ has expectation $\text{E}[\frac{\mathbf{X}^T \cdot \varepsilon}{n}] = 0$ and covariance $\text{Cov}[\frac{\mathbf{X}^T \cdot \varepsilon}{n}] = \frac{\mathbf{X}^T \cdot \mathbf{\Omega} \cdot \mathbf{X}}{n^2}$.
An estimator is called consistent if it *converges in probability*[1] to the underlying population parameter. This implies that the estimator *converges in quadratic mean*. That is, it is asymptotically unbiased and its asymptotic variance vanishes. Given that $\lim_{n\to\infty}(\frac{\mathbf{X}^T \cdot \mathbf{X}}{n})$ is finite and non-singular then $\text{plim}[\widehat{\boldsymbol{\beta}}] = \boldsymbol{\beta}$, that is, $\text{plim}[\frac{\mathbf{X}^T \cdot \varepsilon}{n}] = \mathbf{0}$ because $\text{Cov}[\frac{\mathbf{X}^T \cdot \varepsilon}{n}] = \frac{\mathbf{X}^T \cdot \mathbf{\Omega} \cdot \mathbf{X}}{n^2}$ goes to zero for increasing n due to a finiteness assumption of $\lim_{n\to\infty}(\frac{\mathbf{X}^T \cdot \mathbf{\Omega} \cdot \mathbf{X}}{n})$.

The covariance of the erroneously applied OLS estimator $\widehat{\boldsymbol{\beta}}$ under the presence correlated disturbances is

$$\text{E}[(\widehat{\boldsymbol{\beta}} - \boldsymbol{\beta}) \cdot (\widehat{\boldsymbol{\beta}} - \boldsymbol{\beta})^T \mid \mathbf{\Omega}] = \text{E}[(\mathbf{X}^T\mathbf{X})^{-1}\mathbf{X}^T \varepsilon \cdot \varepsilon^T \mathbf{X}(\mathbf{X}^T\mathbf{X})^{-1}] , \tag{2.24}$$

$$= (\mathbf{X}^T\mathbf{X})^{-1}\mathbf{X}^T \cdot \mathbf{\Omega} \cdot \mathbf{X}(\mathbf{X}^T\mathbf{X})^{-1} . \tag{2.25}$$

Thus it differs from the covariance matrix (2.7) of the OLS estimator $\widehat{\boldsymbol{\beta}}$. However, the direction of this discrepancy can not be given in analytical terms. It might happen that the estimated variance of a single OLS regression parameter β_j, $j \in \{1, \ldots, k\}$, is either smaller or larger than its true value. This has marked consequences for statistical tests and can lead to incorrect statistical conclusions.

2.3   The Impact of Unaccounted Variables

In econometrics serial autocorrelation tests in the regression residuals, like for instance the Durbin and Watson d-statistics, are also used to identify a misspecified regression model in either its functional form or in neglected or unknown but relevant exogenous variables (see Fomby et al., 1984, p 409). With respect to the spatial situation Anselin and Griffith (1988, p 18) state that "a Moran statistic based on the residuals in the misspecified model could potentially indicate significant spatial autocorrelation, even when the error terms are independent." Furthermore they mention (p 19) that "the issue is not necessarily whether expansion [of the model] eliminates spatial autocorrelation in the error terms. Rather, it could be stated in terms of the power of the Moran test against misspecification of the form" of the underlying regression model.

There are numerous examples in the geographical literature which search for exogenous variables either with substantial meaning or of artificial origin which eliminate the spatial autocorrelation in the regression residuals. To mention a few, Boots and Kanaroglou (1988) used as a surrogate the principal component

---

[1] Formally, convergence in probability is defined as $\text{plim}[\widehat{\beta}_{(n)}] \equiv \lim_{n\to\infty} \text{Pr}(|\widehat{\beta}_{(n)} - \beta| > \epsilon) = 0$ where ϵ is an infinitesimal small positive constant.

of the spatial structure to adjust for the spatial autocorrelation; this idea has been extended by Griffith (1996) who used a model driven approach to identify a relevant subset of eigenvectors from the spatial structure; or Haining (1991) who searched in the disease mapping tradition for exogenous variables breaking the spatial ties in the regression residuals.

This section will outline why unknown but relevant variables *can cause* the regression residuals of the remaining model to exhibit spatial autocorrelation. This section takes the perspective that the unknown or neglected variables take the status of a random component because they become part of the disturbances vector and are not observable. This new component has an impact on the covariance of the regression residuals. A formal discussion of this problem is explicated and, furthermore, misconceptions found in the literature are corrected.

2.3.1   The Two Stage Estimation Procedure

As in the previous section first the proper model is presented. This allows me to show the effects of a misspecified model in the next subsection. Let the design matrix $\mathbf{X}$ in the regression model be partitioned into two subsets so that

$$y = \mathbf{X}_1 \cdot \boldsymbol{\beta}_1 + \mathbf{X}_2 \cdot \boldsymbol{\beta}_2 + \varepsilon \tag{2.26}$$

with $\mathbf{X} = [\mathbf{X}_1 \mid \mathbf{X}_2]$ and $\boldsymbol{\beta}^T = [\boldsymbol{\beta}_1^T \mid \boldsymbol{\beta}_2^T]$. Let us assume that we employ a two stage estimation procedure where $\boldsymbol{\beta}_2$ is calibrated first by ordinary least squares. On the resulting regression residuals of this model, which are free from the effects of the design matrix $\mathbf{X}_2$, we apply an OLS regression to obtain an estimator for $\boldsymbol{\beta}_1$, that is,

$$\widehat{\boldsymbol{\beta}}_1 = (\mathbf{X}_1^T \cdot \mathbf{M}_2 \cdot \mathbf{X}_1)^{-1} \cdot \mathbf{X}_1^T \cdot \mathbf{M}_2 \cdot y \tag{2.27}$$

where the idempotent projection matrix $\mathbf{M}_2$ is given by

$$\mathbf{M}_2 = \mathbf{I} - \mathbf{X}_2 \cdot (\mathbf{X}_2^T \cdot \mathbf{X}_2)^{-1} \cdot \mathbf{X}_2^T . \tag{2.28}$$

This derivation is in line with the two-stage estimation in the analysis of covariance (see Hsiao, 1986) where the set of exogenous variables is split in categorical and metric subsets. The estimator $\widehat{\boldsymbol{\beta}}_1$ is again best linear unbiased. In the course of the following discussion only the unbiased property is needed which shall be shown here

$$\mathrm{E}[\widehat{\boldsymbol{\beta}}_1] = \mathrm{E}[(\mathbf{X}_1^T\mathbf{M}_2\mathbf{X}_1)^{-1} \cdot \mathbf{X}_1^T\mathbf{M}_2 \cdot y] \tag{2.29}$$

$$= \mathrm{E}[(\mathbf{X}_1^T\mathbf{M}_2\mathbf{X}_1)^{-1} \cdot \mathbf{X}_1^T\mathbf{M}_2 \cdot (\mathbf{X}_1 \cdot \boldsymbol{\beta}_1 + \mathbf{X}_2 \cdot \boldsymbol{\beta}_2 + \varepsilon)] \tag{2.30}$$

$$= \mathrm{E}[(\mathbf{X}_1^T\mathbf{M}_2\mathbf{X}_1)^{-1} \cdot \mathbf{X}_1^T\mathbf{M}_2 \cdot \mathbf{X}_1 \cdot \boldsymbol{\beta}_1] \tag{2.31}$$

$$\qquad + \mathrm{E}[(\mathbf{X}_1^T\mathbf{M}_2\mathbf{X}_1)^{-1} \cdot \mathbf{X}_1^T\mathbf{M}_2 \cdot \mathbf{X}_2 \cdot \boldsymbol{\beta}_2]$$

$$\qquad + \mathrm{E}[(\mathbf{X}_1^T\mathbf{M}_2\mathbf{X}_1)^{-1} \cdot \mathbf{X}_1^T\mathbf{M}_2 \cdot \varepsilon]$$

$$= \boldsymbol{\beta}_1 + \mathbf{0} + (\mathbf{X}_1^T\mathbf{M}_2\mathbf{X}_1)^{-1} \cdot \mathbf{X}_1^T\mathbf{M}_2 \cdot \mathrm{E}[\varepsilon] \tag{2.32}$$

$$= \boldsymbol{\beta}_1 \tag{2.33}$$

because $\mathbf{M}_2 \cdot \mathbf{X}_2 = \mathbf{0}$ and $\mathrm{E}[\varepsilon] = \mathbf{0}$ by assumption.

2.3.2   The Misspecified Regression Model

Previous results allow us to investigate the effects of a misspecified estimator for β_1 which ignores the impact of the projection matrix $\mathbf{M}_2$ of the now *unknown* partial design matrix $\mathbf{X}_2$. However, before proceeding the partitioned design matrix $\mathbf{X}$ needs to be transformed to ensure that the regression residuals $\widehat{\widehat{\varepsilon}}_1$ in the misspecified model still have an expectation of zero. The partial design matrix $\mathbf{X}_2$ can be decomposed into

$$\mathbf{X}_2 = \mathrm{E}[\mathbf{X}_2] + \mathbf{X}_2^* \tag{2.34}$$

where $\mathrm{E}[\mathbf{X}_2]$ is the matrix of column expectations of $\mathbf{X}_2$. $\mathrm{E}[\mathbf{X}_2]$ has the same row and column dimensions as $\mathbf{X}_2$. The columns of the matrix $\mathbf{X}_2^*$ have zero expectations and measure the variation around the columns of $\mathrm{E}[\mathbf{X}_2]$.

Furthermore, the constant unity vector $\mathbf{1}$ of the partial first design matrix $\mathbf{X}_1$ needs to be augmented by the constant but *unknown* vector $\mathrm{E}[\mathbf{X}_2] \cdot \beta_2$. Except for the parameter estimate related to this new constant term $\mathbf{1} + \mathrm{E}[\mathbf{X}_2] \cdot \beta_2$, this transformation has no consequences on the other parameter estimates. However, the regression residuals $\widehat{\widehat{\varepsilon}}_1$ of the misspecified model will remain orthogonal to the constant term. Thus their expectation stays zero. In fact, the OLS estimation procedure is such that the zero mean of the regression residuals is automatically guaranteed at the cost of potentially biased estimates of the constant term (Kennedy, 1985, p 91).

If the modified first partial design matrix is denoted by $\mathbf{X}_1^*$ then the partitioned regression model is written as

$$\mathbf{y} = \mathbf{X}_1^* \cdot \beta_1 + \mathbf{X}_2^* \cdot \beta_2 + \varepsilon . \tag{2.35}$$

Ad hoc determined, the disturbances of the misspecified model become $\mathbf{X}_2^* \cdot \beta_2 + \varepsilon$. The unknown vector $\mathbf{X}_2^* \cdot \beta_2$ takes now the status of a *random variable* with zero expectation and a specific covariance matrix. The disturbances ε and the new error component $\mathbf{X}_2^* \cdot \beta_2$ are still assumed to be independent from each other. Note that this *additivity* fits well into the generating stochastic mechanism of a normal distributed random variable which assumes that a normal distributed variable is setup by a large sum of diminutive, independent random shocks (see Mood et al., 1974, p 240). This mechanism lays, for instance, the foundation of the central limit theorem. However, generally in empirical research the unknown error component $\mathbf{X}_2^* \cdot \beta_2$ will not be diminutive which can cause severe problems like its dominance of the covariance matrix.

The misspecified estimator for β_1 which ignores the effects of the unknown $\mathbf{X}_2^* \cdot \beta_2$ is given in the standard OLS form by

$$\widehat{\widehat{\beta}}_1 = (\mathbf{X}_1^{*T} \cdot \mathbf{X}_1^*)^{-1} \cdot \mathbf{X}_1^{*T} \cdot \mathbf{y} . \tag{2.36}$$

This estimator is biased as can be seen by following the same derivation as in the previous subsection, i.e.,

$$\mathrm{E}[\widehat{\widehat{\beta}}_1] = \mathrm{E}[(\mathbf{X}_1^{*T} \cdot \mathbf{X}_1^*)^{-1} \cdot \mathbf{X}_1^{*T} \cdot (\mathbf{X}_1^* \cdot \beta_1 + \mathbf{X}_2^* \cdot \beta_2 + \varepsilon)] \tag{2.37}$$

$$= \beta_1 + (\mathbf{X}_1^{*T} \cdot \mathbf{X}_1^*)^{-1} \cdot \mathbf{X}_1^{*T} \cdot \mathbf{X}_2^* \cdot \beta_2 , \tag{2.38}$$

which depends also on the unknown design matrix $\mathbf{X}_2^*$ and the unknown regression parameters β_2. While this result is in-line with the result given in Anselin and Griffith (1988, p 18, see also Anselin 1988, p 126) their derivation is faulty because they claimed that they are using the proper estimator $\widehat{\beta}_1$ and applying "straightforward applications of partitioned inversion". Their approach clearly does not work.

Next the regression residuals $\widehat{\widetilde{\varepsilon}}_1$ of the misspecified model become

$$\widehat{\widetilde{\varepsilon}}_1 = \mathbf{y} - \mathbf{X}_1^* \cdot \widehat{\widetilde{\beta}}_1 \tag{2.39}$$

$$= \mathbf{y} - \mathbf{X}_1^* \cdot (\mathbf{X}_1^{*T} \cdot \mathbf{X}_1^*)^{-1} \cdot \mathbf{X}_1^{*T} \cdot \mathbf{y} \tag{2.40}$$

$$= (\mathbf{I} - \mathbf{X}_1^* \cdot (\mathbf{X}_1^{*T} \cdot \mathbf{X}_1^*)^{-1} \cdot \mathbf{X}_1^{*T}) \cdot (\mathbf{X}_1^* \cdot \beta_1 + \mathbf{X}_2^* \cdot \beta_2 + \varepsilon) \tag{2.41}$$

$$= \mathbf{M}_1 \cdot (\mathbf{X}_2^* \cdot \beta_2 + \varepsilon) \tag{2.42}$$

because $\mathbf{M}_1 \cdot \mathbf{X}_1^* \cdot \beta_1 = 0$ where

$$\mathbf{M}_1 = \mathbf{I} - \mathbf{X}_1^* \cdot (\mathbf{X}_1^{*T} \cdot \mathbf{X}_1^*)^{-1} \cdot \mathbf{X}_1^{*T} \tag{2.43}$$

is a projection matrix in terms of the partial design matrix $\mathbf{X}_1^*$. Equation (2.42) confirms the intuitive composition of the disturbances in a misspecified regression model. Note that there is again a misconception in Anselin and Griffith (1988, p 18) because they switch the definition of their projection matrix "M" from $\mathbf{M}_2$ to $\mathbf{M}_1$ without notifying the reader.

The regression residuals $\widehat{\widetilde{\varepsilon}}_1$ are unbiased, i.e.,

$$\mathrm{E}[\widehat{\widetilde{\varepsilon}}_1] = \mathbf{M}_1 \cdot \mathrm{E}[\mathbf{X}_2^* \cdot \beta_2] + \mathbf{M}_1 \cdot \mathrm{E}[\varepsilon] \tag{2.44}$$

$$= 0 , \tag{2.45}$$

because $\mathbf{X}_2^* \cdot \beta_2$ and ε have zero expectation. However, the covariance between the regression residuals changes substantially

$$\mathrm{E}[\widehat{\widetilde{\varepsilon}}_1 \cdot \widehat{\widetilde{\varepsilon}}_1^T] = \mathbf{M}_1 \cdot \mathrm{E}[\mathbf{X}_2^* \cdot \beta_2 \cdot \beta_2^T \cdot \mathbf{X}_2^{*T}] \cdot \mathbf{M}_1 + \sigma^2 \cdot \mathbf{M}_1 . \tag{2.46}$$

Use has been made here of the fact that ε and $\mathbf{X}_2^* \cdot \beta_2$ are supposed to be independent. This result fits very well into Cressie's observation (1991, p 114) that "one [researcher's] deterministic mean structure may be another [researcher's] correlated error structure" because the unknown component $\mathbf{X}_2^* \cdot \beta_2$ enters the covariance matrix of the regression residuals. Depending on the problem under investigation it may be claimed by a researcher that $\mathbf{X}_2^* \cdot \beta_2$ is not of interest for the mean structure. However it must be dealt with in the covariance matrix. Furthermore, two special cases need to be distinguished here:

1. If the second partial design matrix $\mathbf{X}_2^*$ is irrelevant for explaining the exogenous variable $\mathbf{y}$ then $\beta_2 = 0$. Consequently, the covariance reduces to the standard form of $\mathrm{E}[\widehat{\widetilde{\varepsilon}}_1 \cdot \widehat{\widetilde{\varepsilon}}_1^T] = \sigma^2 \cdot \mathbf{M}_1$. However, the correct (equation 2.27) and the incorrect (equation 2.36) estimators for β_1 are still different (see Fomby et al., 1980, p 403).

2. If the partial design matrices $\mathbf{X}_1^*$ and $\mathbf{X}_2^*$ are mutually orthogonal then the covariance matrix reduces to $\mathrm{E}[\widehat{\bar{\varepsilon}}_1 \cdot \widehat{\bar{\varepsilon}}_1^T] = \mathrm{E}[\mathbf{X}_2^* \cdot \boldsymbol{\beta}_2 \cdot \boldsymbol{\beta}_2^T \cdot \mathbf{X}_2^{*T}] + \sigma^2 \cdot \mathbf{M}_1$. Furthermore, the estimator for $\boldsymbol{\beta}_1$ is unbiased (see Fomby et al., 1980, p 404).

In particular the second case shows that if $\mathbf{X}_2^* \cdot \boldsymbol{\beta}_2$ follows a spatially autocorrelated pattern then a test on spatial autocorrelation in the regression residuals of a model ignoring $\mathbf{X}_2^*$ will indicate significant spatial autocorrelation even if ε is spatially independent. However, in contrast to Anselin and Griffith's (1988, p 18) assertion[2] no statement can be made about the direction of this bias.

---

[2] Anselin and Griffith (1988, p 18) assert that the matrix $\boldsymbol{\beta}_2^T \cdot \mathbf{X}_2^{*T} \cdot \mathbf{M}_1 \cdot \mathbf{W} \cdot \mathbf{M}_1 \cdot \mathbf{X}_2^* \cdot \boldsymbol{\beta}_2$, where $\mathbf{W}$ is a spatial link matrix, is positive. However, this would only be the case if $\mathbf{W}$ is positive definite which it is not.

Chapter 3

The Specification of Spatial Relations

In this chapter I will deal with the quintessence of geographical inquiries which I think distinguishes Geography from other sciences. That is, Geography focuses on the identification, explanation and prediction of processes establishing *spatial relationships* and the *spatial organization* between spatial objects. This chapter borrows on some of the work done by Gatrell (1983) and Jammer (1993) on the nature of space.

The general perspective is that we deal with a set of possible infinite spatial objects. A *force field* determines the physical location or trajectory (e.g., continental drift) of these spatial objects. Thus, this force field establishes a physical order or arrangement of the material objects to keep them in place by mutual local ties. The force field operates *absolutely* in space and time. Furthermore, each object is characterized by a bundle of attributes which are either loosely or firmly attached to it. Energy is needed to overcome the underlying force field. This energy allows spatial objects to interact with each other and to mutually exchange the loosely attached attributes. The way the force field is conquered defines the relationship between the spatial objects as well as their organization which may differ depending on the employed spatial scale and the problem under investigation. These relationships and the spatial organization establish the *relative* space. Usually there is some degree of correlation between absolute and relative space because the energy needed to overcome the force field is related to the strength of the force field.

To build a model of spatial relationships in absolute or relative space, their internal structure needs to be implemented. In absolute space the spatial structure is independent of the attributes of the spatial objects whereas in relative space the structure may as well depend on a comparative evaluation of selected attributes of the spatial objects. In general, spatial structure is a function of the strength of the spatial relationships and/or the strength of the force field which ties the spatial objects together. This function can portray the relationships either as *similarity* or *dissimilarity* between the spatial objects as discussed in details below.

This chapter is organized in two sections. The first section discusses the connections between spatial surfaces, points and areas. Then relations between spatial objects in absolute space are formally introduced. This allows me to outline in absolute space the two specifications of a spatial structures which are topologically[1] similar. These specifications are the *distance* matrix predominately used

---

[1] Strictly mathematically speaking, topology is a branch of geometry describing the properties of a figure that are unaffected by continuous distortions.

in geo-statistics and the binary *contiguity* matrix mostly found in spatial statistics, spatial econometrics, and spatial epidemiology. Usually, a coding scheme is employed onto these specifications of the spatial structure before the data are further processed. The two commonly used coding schemes and a recently introduced variance stabilizing coding scheme are given. Functional relationships in relative space are discussed only in broad terms because of their idiosyncratic character. It is hard to make general statements about their properties. However an example is given in Part III. – The second section will discuss some special spatial structures. These are the regular structures, higher order link matrices and their relation to the distance matrix and the link between local and global link matrices.

3.1   Implementation of Spatial Structures

3.1.1   Typology of Spatial Objects

Let us assume that we are dealing with a two-dimensional bounded study region $\Re$ in which an infinite number of spatial objects is located (see also Laurini and Thompson, 1992, p 175). These objects define via their attributes the spatial surfaces. Clearly, in practical work, we can not handle such an unlimited amount of objects and some sort of abstraction is needed. Ideally we develop a sampling plan and select a set $\mathcal{O}$ of n relevant and representative spatial point objects. For continuous surfaces a discussion on sampling strategies can be found for instance in Cressie (1991), Haining (1990), and Thompson (1992). Less ideally, however, researchers have no influence over the available set $\mathcal{O}$ due to numerous constraints and they have to take the given set of spatial objects for granted. Nevertheless, in contrast to point pattern analysis the position of the sampled spatial objects is by assumption fixed and non-random.

Alternatively, the study region $\Re$ can be partitioned exhaustively into n mutually exclusive areal spatial objects. This partition is called a spatial *tessellation* of the study region $\Re$ into n tiles. Two adjacent areal spatial objects are separated by a common edge. These boundaries may also arise from spatial discontinuity (e.g., political borders or natural barriers). Alternatively, the areas may be created by Voronoi polygons[2] originating from generator points (see for instance Okabe et al., 1994). These generator points could be representative sample point objects taken from the study region $\Re$. Vice versa, reference points can be calculated as centroids from the areal spatial objects. Within a single area the relevant attributes are either integrated measures over its domain or simply based on an assignment from its reference point. It is hoped that an areal object is sufficiently homogenous[3] so that the integrated measure or the

---

[2] The program GAMBINI written by Tiefelsdorf and Boots (1997) can for instance be used to generate multiplicative weighted Voronoi polygons. Commercial GIS packages like for instance Maptitude 4.0 by Caliper Corporations include efficient modules for ordinary Voronoi diagrams.

[3] Violation of the homogeneity assumption is one facet of the modifiable areal unit problem, which assumes that the spatial objects are an aggregation of inhomoge-

simple assignment procedure gives a representative description over its whole enumeration area.

3.1.2   Relations between Spatial Objects

In a strict mathematical sense, spatial relations B are defined on the Cartesian product $\mathcal{O} \times \mathcal{O} = \{(i,j) : i \in \mathcal{O}, j \in \mathcal{O}\}$ of ordered pairs (i,j) of spatial objects. The spatial relations B form a subset of Cartesian product, that is, $B \subseteq \mathcal{O} \times \mathcal{O}$. Irreflexibility is postulated for a spatial relation, which means that a spatial object is not related to itself, that is, $(i,i) \notin B$. Furthermore, if $(i,j) \in B$ and $(j,i) \in B$ for all ordered pairs in the relation, then the relation is said to be symmetric. The concept of spatial relations is more general than Anselin's spatial lag model (1988, Chap. 3), which is borrowed from index shifts in time series analysis.

A set of three spatial objects i, j and k in a symmetric relation B is said to form a *clique* of size three if $(i,j) \in B$ and $(i,k) \in B$ as well as $(j,k) \in B$ (see Cressie, 1991, p 414, and Haining, 1991, p 74 for some examples). This means that in absolute space the three spatial objects i, j, and k are mutual neighbors. A clique is also transitive because from $(i,k) \in B$ and $(j,k) \in B$ follows that $(i,j) \in B$. Tri-valent cliques are observed in most empirical tessellations in the interior of the study region $\mathfrak{R}$.

A relation can be represented either in a binary $n \times n$ link matrix $\mathbf{B}$ or in a spatial graph. If none of the edges in the graph intersect another edge, then the graph is said to be *planar*. Spatial graphs derived from relations in absolute space are always planar. Clearly, these spatial graphs represent *multilateral* and *multidirectional* relations. This differs from temporal relations which allow only sequential links between the past and the present as well as the present and the future, that is, these links are directed along the temporal axis. This multilateral and multidirectional nature of spatial relations complicates spatial analysis matters considerably (Anselin, 1988, p 16).

3.1.3   Similarity Versus Dissimilarity Relations

Figure 3.1 illustrates the topological similarity between areal spatial objects and their reference point objects as well as the specification of the spatial structure as either a graph or a configuration. It relates to an arrangement of the spatial objects in absolute space. First, it starts with a tessellation of the study region $\mathfrak{R}$. Each areal object is separated from its neighboring objects by an edge. Within each areal object lies a reference point. Second, the arrangement of the spatial objects can be specified either as a relational planar graph which connects adjacent spatial objects or as a set of coordinates in Euclidean space. Third, either the graph is coded into a spatial link matrix or the coordinate table is transformed into a distance matrix. Several coding schemes of the spatial

---

neous elemental spatial units (see Openshaw, 1984, Arbia, 1989, and Amrhein and Flowerdew, 1992).

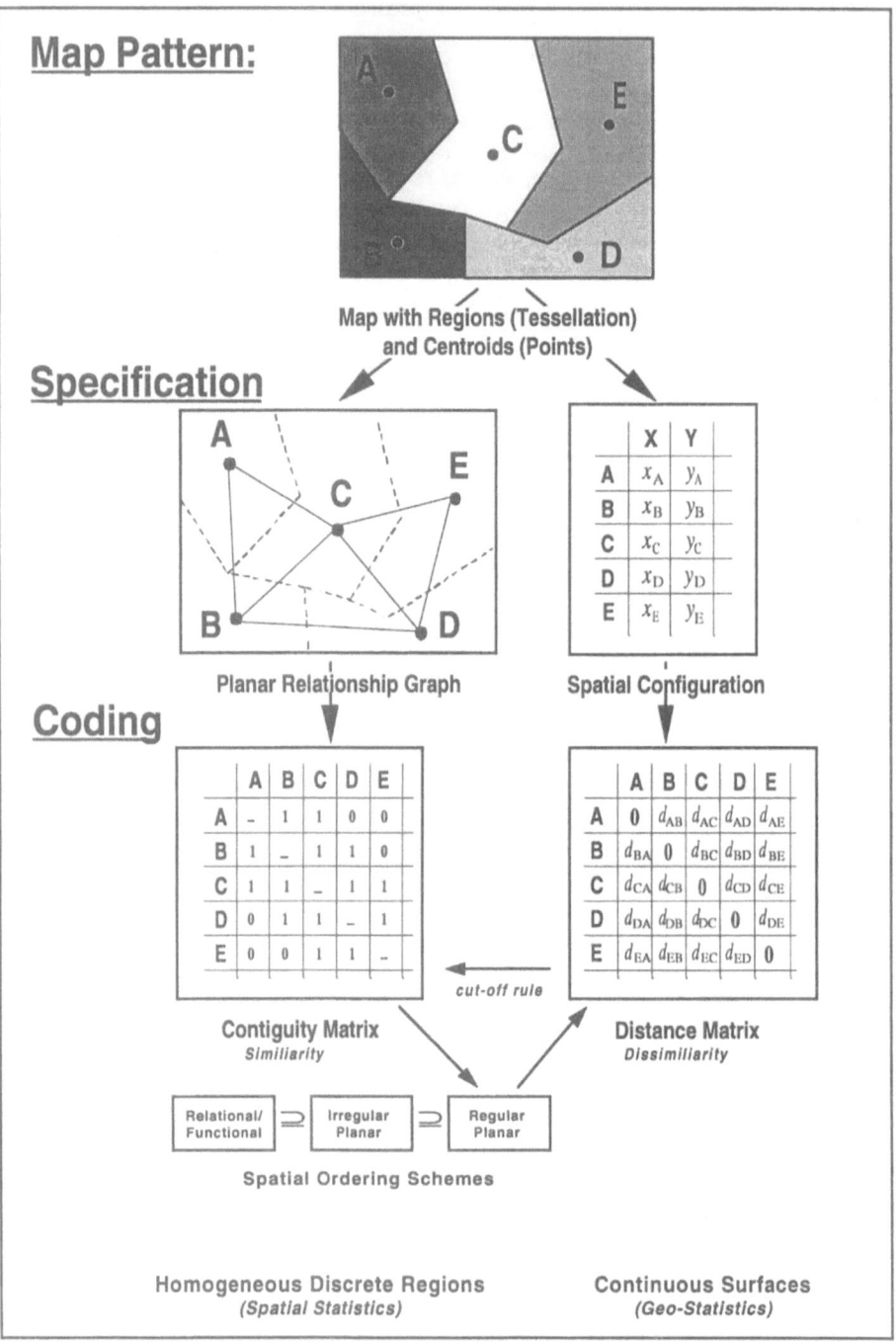

Fig. 3.1. The specification of spatial structures

relations are discussed below. Because both relational matrices are derived from essentially the same information, we can expect that they share some topological similarities. This will be discussed below, too.

While spatial statistics predominately uses the link matrix, geo-statistics focuses on relationships expressed by the distance matrix. In my opinion this distinction has a sound foundation because it leads to different statistical methods. However, presently many new terms have been coined in spatial statistics such as spatial epidemiology or spatial econometrics which relate to specific areas of application. These specialized fields share a common spectrum of employed methods. Thus, I feel that this is an unnecessary fragmentation which prohibits progress instead of generating synergy effects. Furthermore, I do not agree with Anselin's (1988, p 10) statement that the "primary distinguishing characteristic [between spatial statistics and spatial econometrics] seems to be the data-driven orientation of most writings in spatial statistics and the model-driven approach in spatial econometrics". Spatial statisticians are very well aware of models as can be seen by the examples given in Cliff and Ord (1981).

The contiguity matrix measures the similarity between spatial objects. Note that in the literature the terms neighborhood and adjacency are almost used interchangeably for contiguity. Similarities are measured by non-negative numbers. An entry larger than zero indicates that a pair of spatial objects is similar with respect to some attributes, that is for spatial relations in absolute space, they share a common edge. A zero denotes that both spatial objects do not have anything in common. Self-similarity is excluded by the irreflexibility postulate for spatial relations. Usually a similarity matrix is assumed to be symmetric. The spatial contiguity matrix $\mathbf{B}$ is an example of a similarity matrix. Its elements are binary, that is,

$$b_{ij} = \begin{cases} 1 & \text{if } i \text{ and } j \text{ are spatially tied together, and} \\ 0 & \text{otherwise or if } i = j. \end{cases} \qquad (3.1)$$

In contrast, the distance matrix $\mathbf{D}$ measures dissimilarities. Again the elements of a dissimilarity matrix have to be real numbers greater equal than zero. The elements are defined by $d_{ij} = \sqrt{(\vartheta_i - \vartheta_j)^2 + (\omega_i - \omega_j)^2}$ where ϑ and ω are the coordinates of the spatial objects i an j, respectively, in the Cartesian coordinate system.[4] However, the interpretation of its entries is reversed: the larger an entry is the less both associated spatial objects have in common. By definition of the distance matrix a spatial object can not be dissimilar with itself, therefore, the main diagonal has zero entries.

While a transformation from distance to adjacency matrices in absolute space is easily performed for a regular spatial tessellation (see the definition of regular

---

[4] Note that in small scaled maps also the curvature of the earth surface has to be taken into consideration. Furthermore, distance functions in the Minkowski metric $d_{ij}^p = \left((\vartheta_i - \vartheta_j)^p + (\omega_i - \omega_j)^p\right)^{\frac{1}{p}}$ are sometimes used. Note that this metric is only rotational invariant for $p = 2$. The choice of p defines the coding scheme for distance matrices.

tessellations in the next section), a one-to-one reverse transformation from the binary link matrix back to distances can not be performed. This is due to the loss of information based on the measurement scale reduction by moving from the cardinal distance scale to the ordinal characteristic underlying the contiguity matrix. However, as will be shown in the next section there is still a close affinity between both matrices. One possible "cut off rule", transforming a distance matrix $\mathbf{D}$ into a binary contiguity matrix $\mathbf{B}$, is

$$
b_{ij} = \begin{cases} 1 & \text{for } d_L \leq d_{ij} < d_U \text{ and} \\ 0 & \text{otherwise} \end{cases}
\tag{3.2}
$$

where d_L and d_U denote the bounds of a lower and upper distance band, respectively. If $d_L = 0$ then all spatial objects in the vicinity (i.e., radius d_U) of a reference object are selected. Another transformation that inverts the distance matrix into a similarity matrix is $1/d_{ij}^\alpha$. As the exponent α surpasses 3 the transformed distance matrix becomes increasingly similar to the contiguity matrix (see Upton, 1990, p 328). Though distance is measured on a metric scale, often distances beyond a specific threshold are regarded as irrelevant. Consequently, pairs of spatial objects more than this threshold distance apart are considered unrelated. For instance, Wackernagel (1995, p 32) suggests that distances past half the diameter of the study region should be ignored simply because the number of pairs (i, j) beyond this critical distance are decreasing again. Another expression of this irrelevance of higher distances is the presence of a sill in the semi-variogram which is used to characterize spatial dependence in geo-statistics..

3.1.4   Functional Relationships in Relative Space

Often also general relations $\mathcal{G}$ are used which I term *functional* relationships. A mutual weight g_{ij} is assigned to each ordered pair (i, j) in the relation $\mathcal{B}$ so that the elements of the generalized relationship $\mathcal{G}$ become triplets (i, j, g_{ij}).[5] This weight g_{ij} is a function of attributes $\mathbf{x}_i$ and $\mathbf{x}_j$ defined on the spatial objects i and j, that is,

$$
g_{ij} = \begin{cases} g(x_i, x_j) > 0 & \text{if } (i, j) \in \mathcal{B}, \\ 0 & \text{if } (i, j) \notin \mathcal{B} \text{ and } i = j, \end{cases}
\tag{3.3}
$$

for all $(i, j) \in \mathcal{B}$. General relations $\mathcal{G}$ can be expressed again in matrix notation by $\mathbf{G}$. A relation $\mathcal{B}$ is a sub-class of a functional relationship $\mathcal{G}$ and can be deduced from it by applying an indicator function on the weights g_{ij}. In spatial investigations, functional relationships define also a measurement space into which the spatial objects are embedded (Gatrell, 1983).

---

[5] Notice that formally distances can be treated as generalized relations. Here the triplet becomes (i, j, d_{ij}).

In relative space, functional relationships still can be visualized by graphs which are defined by a system of exchange mechanisms between the spatial objects. Nevertheless, the edges of the graph no longer need to represent the spatial arrangement of the objects. Consequently, the edges are allowed to intersect. Furthermore, they can be weighted by some attributes. Also the edges can be directed, thus, the functional relationship g_{ij} between the ordered pair objects (i,j) need not be the same as g_{ji} for the ordered pair (j,i). However, a non-symmetric functional relationships matrix $\mathbf{G}^*$ can be transformed to symmetry by

$$\mathbf{G} = \frac{1}{2}(\mathbf{G}^* + \mathbf{G}^{*T}) \, . \tag{3.4}$$

Note that the autocorrelation statistic Moran's I is invariant under this symmetry transformation (see Cliff and Ord, 1981, p 18 and p 202).

Usually the links in relative space represent some sort of interaction potential between the spatial objects which are expressed as similarities. As for the distance relationships also for graphs in relative space, smart cut off rules can be applied to simplify the underlying graph and to identify key spatial objects and their linkages in the spatial system (Haining, 1990, p 73). Methods for achieving this reduction are described in Nystuen and Dacey (1961), Holmes and Haggett (1977), and Gatrell (1979).

3.1.5   Coding Schemes of Spatial Structure Matrices

It is common practice in spatial statistics, geo-statistics and spatial econometrics to convert the general spatial link matrix $\mathbf{G}$ and the binary spatial link matrix $\mathbf{B}$ by using *coding schemes*. This is done with the intention of coping with the heterogeneity which is induced by the different linkage degrees of the spatial objects. The linkage degree of a single spatial object i is defined by the total sum of its interconnections with all other spatial objects, that is,

$$d_i = \sum_{j=1}^{n} g_{1j} \tag{3.5}$$

These coding schemes are:

Definition 4 (The globally standardized C-coding scheme). *Let D be the overall connectivity of elements in* $\mathbf{G}$, *i.e.,* $D \equiv \sum_{i=1}^{n} \sum_{j=1}^{n} g_{ij}$. *Then*

$$\mathbf{C} \equiv \frac{n}{D} \cdot \mathbf{G} \, . \tag{3.6}$$

Definition 5 (The row-sum standardized W-coding scheme). *Let the vector* $\mathbf{d}$ *define the* local *linkage degrees of the spatial objects by the row-sums of* $\mathbf{G}$, *i.e.,*

$$\mathbf{d} \equiv \left(\sum_{j=1}^{n} g_{1j}, \sum_{j=1}^{n} g_{2j}, \ldots, \sum_{j=1}^{n} g_{nj} \right)^T \, . \tag{3.7}$$

If we are dealing with the binary link matrix **B** *then the elements of* **d** *measure the number of spatial neighbors. Then*

$$\mathbf{W} \equiv [\mathrm{diag}(\mathbf{d})]^{-1} \cdot \mathbf{G} . \qquad (3.8)$$

In addition, Tiefelsdorf, Griffith and Boots (1999) introduce a new variance stabilizing coding scheme:

Definition 6 (The variance stabilizing S-coding scheme). *Let the vector* **q** *be defined by*

$$\mathbf{q} \equiv \left(\sqrt{\sum_{j=1}^{n} g_{1j}^2}, \sqrt{\sum_{j=1}^{n} g_{2j}^2}, \dots, \sqrt{\sum_{j=1}^{n} g_{nj}^2} \right)^T . \qquad (3.9)$$

The preliminary link matrix **S*** *is given by* $\mathbf{S}^* \equiv [\mathrm{diag}(\mathbf{q})]^{-1} \cdot \mathbf{G}$. *From* **S*** *the overall linkage degree is calculated by* $Q \equiv \sum_{i=1}^{n} \sum_{j=1}^{n} s_{ij}^*$ *so that* **S*** *is finally scaled by*

$$\mathbf{S} \equiv \frac{n}{Q} \cdot \mathbf{S}^* . \qquad (3.10)$$

For a binary link matrix **B** *the definition of* **S** *simplifies substantially:*

$$\mathbf{S} \equiv \frac{n}{\sum_{i=1}^{n} \sqrt{d_i}} \cdot [\mathrm{diag}(\mathbf{d})]^{-\frac{1}{2}} \cdot \mathbf{B} . \qquad (3.11)$$

Note that the novel spatial link matrix **S** is as easily calculated as its traditional counterparts **C** and **W**. All coding schemes satisfy the *common constraint* that the overall sum of their elements equals n. This puts autocorrelation statistics derived from them roughly on a common scale.

Introduction of the novel S-coding scheme is motivated by the fact that it stabilizes the *topology induced heterogeneity* in the variances of local Moran's I_i. This topology induced heterogeneity can have a pronounced impact on the outcome of spatial autocorrelation tests and estimations.[6] The properties of the S-coding scheme lie in between those of the row-sum standardized W-coding scheme and the globally standardized C-coding scheme. For moving average processes, the local variance on the main-diagonal of the spatial co-variance matrix $\boldsymbol{\Omega}$ becomes constant, while the heterogeneity in the local variances is alleviated for autoregressive spatial processes (see Tiefelsdorf, Griffith and Boots, 1999). The topology induced heterogeneity stems from the variation in the local linkage degrees (for example see Haining, 1990, Figs. 3.6 and 3.8). Identical variances are a necessary condition for a spatial process to be stationary.

---

[6] The topology induced heterogeneity is also related to edge effects for bounded study regions $\boldsymbol{\Re}$: The C-coding scheme emphasizes spatial objects with a relatively large number of connections, such as those in the interior of a study region. In contrast, the W-coding scheme assigns higher leverage to spatial objects with few connections, such as those on the periphery of the study region.

To simplify our subsequent discussion notationally let us define a placeholder link matrix $\mathbf{V}$ which stands either for

$$\mathbf{V} \equiv \begin{cases} \mathbf{C} & \text{for the } C\text{-coding scheme, or} \\ \mathbf{S} & \text{for the } S\text{-coding scheme, or} \\ \mathbf{W} & \text{for the } W\text{-coding scheme.} \end{cases} \tag{3.12}$$

The coded spatial link matrix $\mathbf{V}$ is not symmetric for the S- and W-coding scheme.

3.2   Some Special Spatial Structures

This section will deal with some special spatial link matrices defined on adjacency relations. Thus attention is restricted primarily to the binary relation matrix $\mathbf{B}$ in absolute space. First the special class of regular tessellations is introduced. Then the higher order link matrices are discussed. These allow me to show a loose link between the binary relation matrix $\mathbf{B}$ and the distance matrix $\mathbf{D}$. Finally, the global link matrix $\mathbf{B}$ will be decomposed into its local components $\mathbf{B}_i$.

3.2.1   Regular Spatial Structures

Regular spatial structures are characterized by the fact that all their edges are of equal length and that the inner angles between the edges are also identical. The regular structures are formed from first order equidistant generator points. Their tessellations consist of convex tiles. Figure 3.2 gives a visual example of the only three available regular spatial structure, which are, triangles, rectangles and hexagons, here with 256 tiles each. The triangular and rectangular arrangements of triangles both share the same overall linkage degree. However, they differ in their geometrical symmetry properties, in their shapes and in their diameters, which measures the largest distance from boundary to boundary within the tessellation. These differences can have an impact on the results of spatial autocorrelation analysis (see Chapter 8.2.2).

Notice that vertices within the study regions $\mathfrak{R}$ of triangular and rectangular tessellations are not trivalent. This leads to some ambiguity in the determination of the spatial neighbors. A vertex can be interpreted as a border of zero length. Thus spatial neighbors can be defined either on a common edge or on a common vertex or on both. In analogy to moves in chess, the first definition is called *rook's* case, whereas the second definition is called *bishop's* case and the last definition is termed *queen's* case. Clearly the bishop's and queen's case lead to non-planar spatial relation graphs. For this reason attention here is focused solely on the rook's case. Notice further that the triangles and rectangles are unstable tessellations. The slightest translation of their arrangement changes the topology relation. In contrast, due to the dovetail arrangement of the cells, the hexagonal tessellation is extremely stable against distortions. Boots (1977)

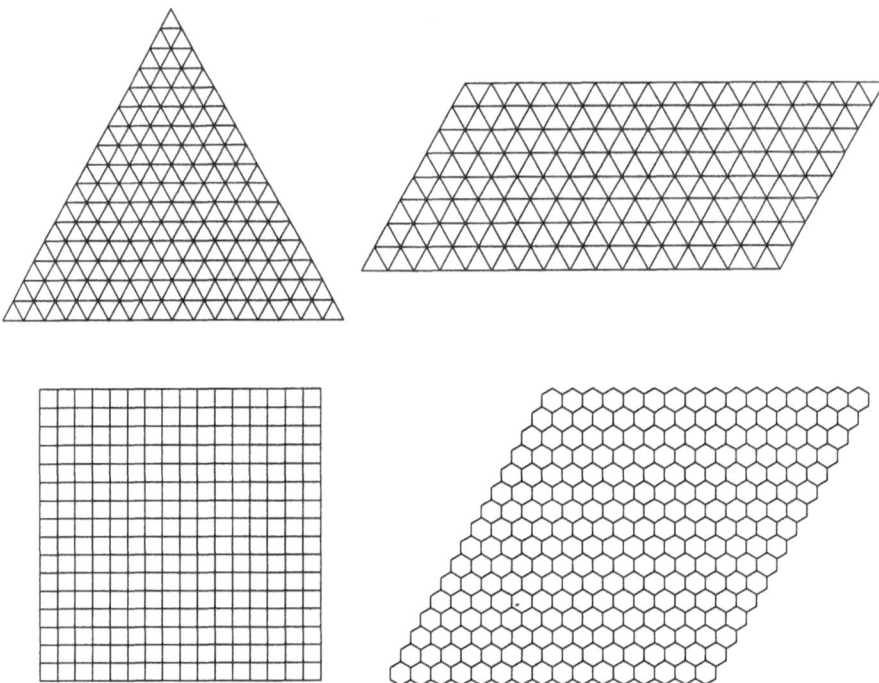

Fig. 3.2. Two triangular tessellations arranged in either a triangle or a rectangle, a rectangular tessellation, and a hexagonal regular tessellation with 256 spatial objects

and Okabe and Sadahiro (1997, p 1549) notice that most random and irregular planar tessellations vary around an average hexagonal partition of the study region $\Re$ (i.e., the tiles in the interior have on average 6 neighbors).

3.2.2   Higher Order Link Matrices

From the binary spatial link matrix $\mathbf{B}$ defined on the first order spatial neighborhood relation between adjacent spatial objects, higher order link matrices can be calculated. It is known from graph theory (see for instance Tinkler, 1977) that the component $b_{ij}^{(p)}$ of a binary spatial link matrix in the p-th power, that is, $\mathbf{B}^p$, measures the number of paths to move in exactly p steps from spatial object i to object j. The first time $b_{ij}^{(p)}$ becomes greater than zero indicates that the spatial objects i and j are exactly p steps apart along the spatial graph. Note that it is not guaranteed that if $b_{ij}^{(p_L)} \geq 1$ then also $b_{ij}^{(p_U)} \geq 1$ for $p_U > p_L$. In fact, it might become zero again if between the spatial objects i and j no path with p_U intermediate steps exists. Thus the spatial link matrix $\mathbf{B}$ has to be powered up successively and checked at each step to determine the set of spatial objects which are p or less steps apart. The value of p for which each component $b_{ij}^{(p)}$ has become at least once greater than zero is called the diameter of the graph and is denoted by $p_{\max}$. It measures the shortest path between the

furthest apart spatial objects. Technical details on a numerically efficient way
of calculating higher order link matrices for large tessellations can be found in
Anselin and Smirnov (1996).

Higher order links are often used as substitute for spatial lags in spatial
correlogram analysis (see Arbia, 1989, Cliff et al., 1975, Cliff and Ord, 1981,
and Oden, 1984). While in time series analysis the notion of serial lags is well
defined because the observations are equally spaced along the unidirectional
temporal axis, in spatial statistics there are no evenly spaced distance bands
even for regular spatial structures in the rook's case. For instance, the Euclidean
distances in a rectangular arrangement of spatial objects at lag order 2 (crossing
two edges) are $\sqrt{2}$ and 2 units for intermediate directions as well as for the prime
directions, respectively. Furthermore, Anselin (1988, pp 23–26) argues forcefully,
that spatial lags must not be interpreted as unidirectional shifts but have more
in common with the distributed lag concept in econometrics. These relate a
measurement at one spatial object to a set of measurements on other spatial
objects in the study region $\mathfrak{R}$.

3.2.3   A Link between the Contiguity Matrices and Spatial Configurations

For planar spatial tessellations the configuration of the representative points is
a two-dimensional $n \times 2$ matrix (see Figure 3.1). Thus the distance matrix $\mathbf{D}$
can have at the most rank 2. The question which arises now is, are we able to
retrieve from the distance matrix the configuration of the spatial objects? Up to
a translation and a rotation we are in fact able to do so in Euclidean space by
using the Young and Householder theorem (Young and Householder, 1938). The
basic relation is given by the transformation

$$\mathbf{D}_c = \mathbf{M}_{(1)} \cdot \mathbf{D}^{(2)} \cdot \mathbf{M}_{(1)} \tag{3.13}$$

where $\mathbf{D}^{(2)}$ is the distance matrix whose elements are d_{ij}^2 (note that $\mathbf{D}^{(2)} \neq \mathbf{D}^2 = \mathbf{D} \cdot \mathbf{D}$). The projection matrix $\mathbf{M}_{(1)}$ has been defined in equation (2.16)
and $\mathbf{D}_c$ is the so-called doubly centered distance matrix (for details see Borg
and Lingoes (1987, Chap. 18) or Upton and Fingleton (1989, pp 173–177)). The
matrix $\mathbf{D}_c$ is positive semi-definite,[7] thus it has two eigenvalues greater than
zero and the remaining eigenvalues are zero. The two eigenvectors associated
with the non-zero eigenvalues mirror the configuration of the spatial objects.

To derive a configuration of the spatial objects from the contiguity matrix
we have to start from the binary spatial link matrix $\mathbf{B}$. In a *first* step a transformation has to be found so that the binary spatial link matrix resembles a

---

[7] Note that the distance matrix $\mathbf{D}^{(2)}$ is indefinite because its trace equals zero. Thus it
has three eigenvalues unequal to zero which, however, have to add up to zero. Two of
these eigenvalues and their related eigenvectors are identical to those of $\mathbf{D}_c$ whereas
the principal eigenvector related to the first eigenvalue measures the distance of
the spatial objects from the origin of the configuration. Therefore, it is a centrality
measure which can be calculated also from the configuration.

distance matrix. Then, in a *second* step, the Young and Householder theorem can be used to obtain the configuration of the spatial objects. Let us define a scalar indicator function h on the components of the higher order spatial link matrix $\mathbf{B}^p$ such that

$$h(b_{ij}^{(p)}) = \begin{cases} 1 & \text{if } b_{ij}^{(p)} \geq 1 \text{ for the first time and} \\ 0 & \text{otherwise.} \end{cases} \tag{3.14}$$

This transformation allows us to build from the sequence of higher order link matrices $\mathbf{B}^p$ a matrix $\mathbf{B}_D$ by applying

$$\mathbf{B}_D = \sum_{p=1}^{p_{\max}} p \cdot h(\mathbf{B}^p) . \tag{3.15}$$

An example of such an order of spatial lags matrix can be found in Cliff et al. (1975, Table 6.1). The matrix $\mathbf{B}_D$ is a dissimilarity matrix and its components indicate how many shortest path steps apart the pairs of spatial objects are. Thus it resembles the spatial distance matrix. However, $\mathbf{B}_D$ is based on an ordinal scale whereas the Young and Householder theorem requires a metric Euclidean scale. Consequently, $\mathbf{M}_{(1)} \cdot \mathbf{B}_D^{(2)} \cdot \mathbf{M}_{(1)}$ will not be positive semi-definite. Nevertheless, it can be expected that it has only two dominant positive eigenvalues, whose related eigenvectors reflect the configuration of the spatial objects adequately. Initial experiments with regular spatial structure support this suspicion, but clearly more research is needed here.

3.2.4   Local Spatial Structures

The global spatial link matrix $\mathbf{G}$ can be decomposed into n local link matrices $\mathbf{G}_i$ which are symmetric if $\mathbf{G}$ is symmetric. For the i-th spatial object let the local link matrix $\mathbf{G}_i$ be defined by

$$\mathbf{G}_i \equiv \begin{pmatrix} 0 & \cdots & 0 & g_{1i} & 0 & \cdots & 0 \\ \vdots & \ddots & \vdots & \vdots & \vdots & \ddots & \vdots \\ 0 & \cdots & 0 & g_{i-1,i} & 0 & \cdots & 0 \\ g_{i1} & \cdots & g_{i,i-1} & 0 & g_{i,i+1} & \cdots & g_{in} \\ 0 & \cdots & 0 & g_{i+1,i} & 0 & \cdots & 0 \\ \vdots & \ddots & \vdots & \vdots & \vdots & \ddots & \vdots \\ 0 & \cdots & 0 & g_{ni} & 0 & \cdots & 0 \end{pmatrix} . \tag{3.16}$$

Note the star-shaped arrangement of the components in $\mathbf{G}_i$ with the reference spatial object i in its center and the blocks of structural zeros surrounding it. This allows us to express the global link matrix as $\mathbf{G} = \frac{1}{2} \sum_{i=1}^n \mathbf{G}_i$. The factor $\frac{1}{2}$ emerges because each component g_{ij} in the summation is counted twice. The same decomposition holds for the binary spatial link matrix $\mathbf{B}$. This decomposition was first introduced by Tiefelsdorf and Boots (1997) in their work on local Moran's I_i.

The spatial objects $j \neq i$ are related as satellites to the reference object i if $g_{ij} \neq 0$. Because these satellites are not related to each other and do not impose a mutual order relationship the local link matrix $\mathbf{G}_i$ is aspatial. The spatial configuration can not be derived from it with the method outlined in the previous subsection. The spectrum of eigenvalues of $\mathbf{G}_i$ can be given in analytical terms by

$$\left\{-\sqrt{\textstyle\sum_{j=1}^{n} g_{ij}^2},\, 0,\ldots,0,\, \sqrt{\textstyle\sum_{j=1}^{n} g_{ij}^2}\right\} . \tag{3.17}$$

Note that only two eigenvalues are non-zero which is again an indication of the aspatial character of the local link matrix $\mathbf{G}_i$.

The additive composition of the global link matrix $\mathbf{G}$ can extended for the coded link matrices $\mathbf{V}$. In general terms it holds that $\frac{1}{2} \cdot (\mathbf{V} + \mathbf{V}^T) = \sum_{i=1}^{n} s_i \cdot \mathbf{G}_i$ with s_i being defined as

$$s_i = \begin{cases} \frac{n^2}{2 \cdot D} & \text{for the } C\text{-coding scheme,} \\[2mm] \frac{n}{2 \cdot d_i} & \text{for the } W\text{-coding scheme, and} \\[2mm] \frac{n^2}{2 \cdot Q \cdot q_i} & \text{for the } S\text{-coding scheme.} \end{cases} \tag{3.18}$$

A derivation of these scaling coefficients s_i can be found in Tiefelsdorf and Boots (1997) and Tiefelsdorf, Griffith and Boots (1999). Note that for the C-coding scheme s_i is constant across the local link.

Chapter 4

Gaussian Spatial Processes

While Chapter 2 focused on the merger of the stochastic disturbances with a set of exogenous variables by means of an ordinary linear regression model this chapter will deal with the link between the disturbances and the exogenously defined spatial structure. This link and the assumption that the disturbances follow a normal distribution define a Gaussian spatial process. This chapter seeks for statistical tests allowing one to identify the underlying spatial and functional relationships which tie the spatial objects stochastically together. The magnitude of the test statistics can be regarded as an indicator for the strength of these relationships.

The first section presents the merger of the disturbances and the spatial structure. Furthermore, some general properties of and assumptions on spatial processes are discussed. The next section makes the explicit assumption that the disturbances follow a normal distribution. This allows me to fully specify autoregressive and moving average spatial Gaussian processes in a regression model. Also the specific properties of these Gaussian spatial processes are given in this section. The concluding section presents some commonly used test principles and their related test statistics for Gaussian spatial processes.

4.1 Blending the Disturbances with a Spatial Structure

A stochastic process can be represented as a *black box*. Inside this black box operates an unknown mechanism which transforms some *random input* into an *output* which is observable. McCleary and Hay (1980) present an excellent analogy of this unknown stochastic mechanism in the context of time series analysis. They describe a coffee brewing process $f(\cdot)$, which uses an electric coffee machine, to introduce the notion of autoregressive processes and moving average processes. The input to the t-th brewing experiment at time t is an amount of water ε_t which is unknown, randomly and independently selected from experiment to experiment. The coffee machine is regarded as a black box processing the water fed to it between times t and $t+1$. The observed quantity of coffee produced by the t-th brewing process is the output y_{t+1} at time $t+1$. Note that it takes some time to process the coffee (i.e., until we can observe the output from a given input). This experiment is repeated n times at fixed time intervals. Now the coffee machine might withhold some water in between the runs of the experiment and releases this retained water in a later run. The amount of retained water might depend on the amount of water fed to it and/or the output of coffee produced. Thus the temporal linked process $f(\varepsilon_t) \to y_{t+1}$ is out of alignment and might depend to a lesser extend also on the previous input

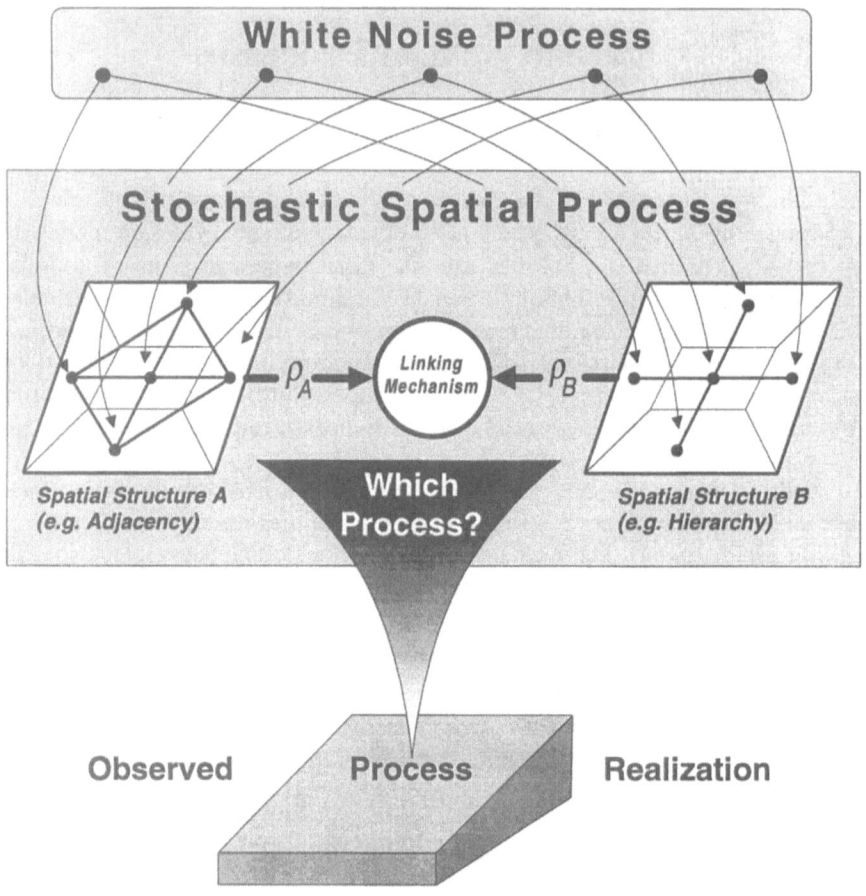

Fig. 4.1. The components and working mechanism of a stochastic spatial process

of water ε_{t-1} and previous output of coffee y_t, i.e., $f(\varepsilon_t, \varepsilon_{t-1}, y_t) \rightarrow y_{t+1}$. The partial dependence of the contemporary output y_{t+1} on the previous input ε_{t-1} to the brewing process gives rise to a *moving average process*. While the partial dependence of the contemporary output y_{t+1} on itself, that is, the previous output y_t, is termed an *autoregressive process*.

4.1.1 The Internal Composition of Stochastic Spatial Processes

To develop a model of spatial processes we need to analyze their internal structure. Figure 4.1 depicts the internal structure of a spatial process. In statistical terms the most general spatial process states that the probability of observing a specific outcome of the response variable y_i at the spatial object i depends on the realizations of the spatial objects at all other locations, i.e.,

$$
\begin{aligned}
&\Pr(Y_i \le y_i \mid Y_1 \le y_1, \ldots, Y_{i-1} \le y_{i-1}, Y_{i+1} \le y_{i+1}, \ldots, Y_n \le y_n) \\
&= F(y_i \mid y_1, \ldots, y_{i-1}, y_{i+1}, \ldots, y_n) .
\end{aligned}
\tag{4.1}
$$

Clearly this indiscriminate dependency is not very helpful because it does not provide useful insight into the specific relationships between the spatial objects. Or following loosely Tobler's (1970) "first law of geography: everything is related to everything else, but specific things are more related than other things". It is precisely the spatial structure, which channelizes this mutual dependence of the spatial objects and allows one to formulate specific hypotheses on the spatial process under consideration. From the perspective of a single spatial object this dependence can be written as

$$
\Pr(Y_i \le y_i \mid Y_j \le y_j, \forall \, j \text{ such that } v_{ij} > 0)
\tag{4.2}
$$

where "j such that $v_{ij} > 0$" denotes that a spatial object j is related to spatial object i and v_{ij} is a component of the coded spatial link matrix $\mathbf{V}$. This can be for instance the first order neighbors of the spatial object i. Consequently, the spatial structure is the key to any spatial process.

Usually the spatial structure and the stochastic mechanism operating within a spatial process are unknown. The same applies also, at least partially, to the input of a spatial process which drives the spatial process. The input may be a pure random vector or a vector composed of deterministic and random components like in regression analysis. If the components of the random vector are independently identical normal distributed the input of the stochastic process is called *white noise*[1] and leads to Gaussian processes. Within the spatial process these random input components are tied together by the links of a spatial structure. The strength of this linking mechanism is determined by the autocorrelation level ρ whereas the linking mechanism itself can take several forms, such as a conditional or simultaneous autoregressive, a moving average or a spatial lag in the exogenous variable specification. If an autocorrelation level ρ associated with a spatial structure is zero then this spatial structure is irrelevant, i.e.,

$$
\Pr(Y_i \le y_i) = \Pr(Y_i \le y_i \mid Y_j \le y_j, \forall \, j \text{ such that } v_{ij} > 0) .
\tag{4.3}
$$

Together these internal components of the spatial process generate the observed map pattern. To make a spatial process operational a temporal perspective has to be taken (i.e., its input has to precede its output). This differentiates a spatial process, whose structure is defined by an underlying theory (see for instance, Stetzer, 1982, or Molho, 1995), from the misspecification perspective employed in Chapter 2.3. The logical structure of this temporal perspective is shown for instance in Haggett's time-space "cone" (see Cliff and Ord, 1981, p 10). In Part III I will present another example on the theoretically guided process perspective by linking a spatial process to an interaction process.

---

[1] The term white noise comes from spectral analysis and denotes white light. Its spectra, i.e., composition, is constant over the whole visible frequency band.

4.1.2   Constraining Assumptions on Spatial Stochastic Processes

The key problem in identifying a spatial process is that we can not conclude un-equivocally which spatial structure and which specification of the spatial process have generated the observed output. This uncertainty stems from the random input into the spatial process. For instance, depending on the specific realiza-tion of the white noise input vector the observed map pattern may appear to be more in line with the spatial structure $\mathcal{B}$ than with the spatial structure $\mathcal{A}$, which defines the underlying spatial process. The situation is further com-plicated, because a mixture of both spatial structures may be involved in the generation of the observed map pattern. Clearly some statistical assumptions are needed which allow us to estimate with statistical certainty the underlying spatial process from the single observed map pattern. These will be outlined below. A procedure to identify statistically the underlying spatial structure is given in Part III.

Having only one realization of the spatial process at hand is not an opera-tional situation. Thus some stability assumptions are needed to generalize char-acteristics of this process realization to the whole spatial process. The statistical concepts of stationarity, isotropy and ergodicity delineate the stability assump-tions. In basic terms the underlying idea of these statistical constructs is that the one observed map pattern can be regarded as a composition of several independ-ent sub-samples. To discuss the *stationarity* and *isotropy* assumption we need the joint-distribution function of the spatial process which can be expressed by means of the conditional distribution given in equation (4.1) as

$$
\begin{aligned}
F(y_1, &\ldots, y_i, \ldots, y_n) \\
&= F(y_i \mid y_1, \ldots, y_{i-1}, y_{i+1}, \ldots, y_n) \cdot F(y_1, \ldots, y_{i-1}, y_{i+1}, \ldots, y_n) \ .
\end{aligned}
\tag{4.4}
$$

Imagine that the spatial objects $1, \ldots, n$ are located within a window of a larger study region and that the spatial structure does not change by translations and rotations of this window. Then, if distribution $F(\cdot)$ does not change under a translation of the window, the process is called stationary and if, furthermore, the distribution does not change under a rotation of the window the process is la-belled isotropic. Consequently, all moments of the distribution $F(\cdot)$ are invariant under these translations and rotations. An isotropic and stationary spatial pro-cess works uniformly over space and time and reflects that its generating mech-anism has reached a statistical equilibrium, that is, a constant joint distribution function. The *ergodicity* constraint states that if we have two non-intersecting windows far enough apart from each other then the distributions of the response variable in both windows are independent from each other.

In combination stationarity, isotropy, ergodicity constraints parallel the clas-sical sampling assumptions, that are, an equal distribution of all observations and independence of the observations, and allow one to characterize the underlying spatial process by just one realization. The statistical theory underlying these concepts of stationarity and ergodicity are much more subtle than presented here and a detailed discussion is beyond this monograph. Good introductory

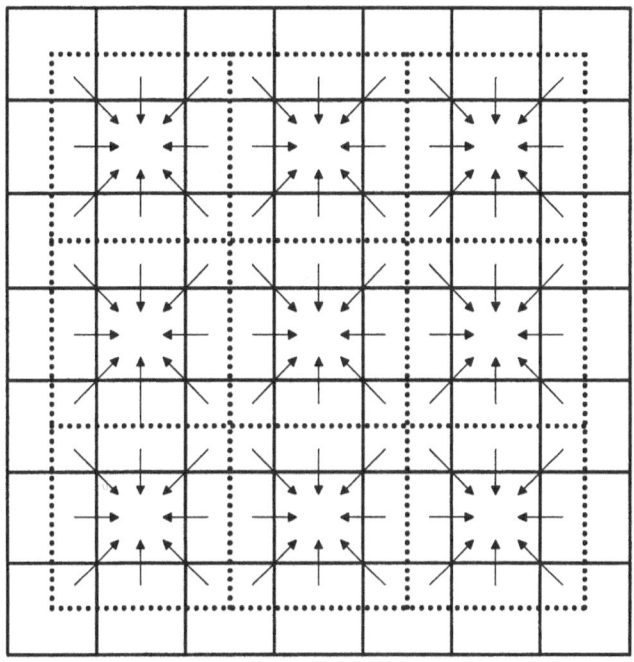

Fig. 4.2. Non-hierarchical aggregation of elemental spatial objects defined by a rect-angular 7×7 tessellation into a rectangular 3×3 tessellation

discussions regarding spatially distributed phenomena can be found in Anselin (1988, Chap. 5), Cressie (1991, Sect. 2.3) and Okabe, Boots, and Sugihara (1992, Sect. 1.3.8).

4.1.3   Excursion: Non-Hierarchical Aggregation of Spatial Objects

So far two causes for autocorrelation between the spatial objects have been introduced. These were a misspecified regression model or an underlying spatial process. However, the observed spatial objects may also be tied together by non-hierarchical aggregation of the elemental spatial objects (see Anselin, 1988, p 12). Often in empirical investigations the elemental spatial objects are unknown and the used tessellation has no affinity with the underlying data generating process. This non-hierarchical aggregation splits the elemental spatial objects and partially assigns them into new aggregated spatial objects. The elemental spatial objects are supposed to be homogeneous whereas the aggregates become non-homogeneous because they are composed of different objects. Furthermore, adjacent aggregates must be correlated because they share attributes from the divided elemental spatial objects.

Figure 4.2 displays such a spatial aggregation procedure. The tiles enclosed by a solid lines are the elemental spatial objects. The tiles with the dotted border lines are the aggregates. Because the borders of aggregates transect the elemental spatial objects their attributes are assigned proportionally to the underlying area to the aggregates. The arrows in Figure 4.2 depict this assignment procedure. Most Geographic Information Systems can perform this aggregation in their overlay modules. Note that this non-hierarchical aggregation of the elemental spatial object will lead to positive spatial autocorrelation. The same applies to a misspecified regression model if the omitted variables operate on a larger spatial scale.

4.2   Definition and Properties of Gaussian Spatial Processes

While preceding sections discussed stochastic spatial processes in general terms this section gives their explicit forms for the special class of Gaussian spatial processes. Furthermore, the properties of the class members are outlined and compared against each other. The class of Gaussian spatial processes is based on normal distributed regression disturbances, that is, white noise input. Consequently, a Gaussian spatial process is parameterized by two groups of parameters (i.e., the expected vector of responses for the n spatial objects and their mutual covariance matrix denoting the interaction of the $n \times n$ pairs of spatial objects, respectively). For a normal distribution, because the joint distribution in equation (4.4) is fully characterized by these two sets of parameters, it is strictly stationarity[2] (Haining, 1990, p 67).

All these Gaussian processes share the same general structure of their joint-density function for the observed realizations y_i at the spatial objects $i \in \{1, \ldots, n\}$, i.e.,

$$f(y_1, \ldots, y_n) = (2\pi)^{-\frac{n}{2}} \cdot |\mathbf{\Omega}(\rho)|^{-\frac{1}{2}}$$
$$\cdot \exp\left(\frac{1}{2} \cdot (\mathbf{y} - \mathbf{X}\boldsymbol{\beta})^T \cdot \mathbf{\Omega}^{-1}(\rho) \cdot (\mathbf{y} - \mathbf{X}\boldsymbol{\beta})\right) \tag{4.5}$$

where disturbances are defined by the vector $\varepsilon = \mathbf{y} - \mathbf{X}\boldsymbol{\beta}$. The term $|\mathbf{\Omega}(\rho)|^{-\frac{1}{2}}$ is called the "Jacobian" which guarantees that the integral over the domain of the density function (4.5) equals unity. The expectation of the endogenous variable is $E(\mathbf{y}) = \mathbf{X}\boldsymbol{\beta}$. The covariance $Cov(\varepsilon \cdot \varepsilon^T) = E(\varepsilon \cdot \varepsilon^T)$ between the spatial objects is characterized by the covariance matrix $\mathbf{\Omega}(\rho)$ which depends on a spatial autocorrelation parameter ρ. Thus disturbances are distributed as $\varepsilon \sim N(0, \mathbf{\Omega}(\rho))$. The disturbances ε can be expressed in terms of an independent, identical normal distributed vector $\boldsymbol{\eta} \sim N(0, \sigma^2 \cdot \mathbf{I})$ by the transformation

$$\varepsilon = \boldsymbol{\eta} \cdot \mathbf{\Omega}^{\frac{1}{2}}(\rho) . \tag{4.6}$$

---

[2] In contrast to *weak stationarity*, which refers only to the expectation and variance of a stochastic process, *strong stationarity* refers to all higher orders of moments (i.e., to the full-fledged distribution given in equation (4.4)).

By treating the variance σ^2 of η as constant over all spatial objects, potential heteroscedasticity is discarded.[3] What differentiates the spatial processes is the explicit specification of the co-variance matrix $\Omega(\rho)$. Recall that a fundamental property of co-variance matrices is their positive definiteness (i.e., all their eigenvalues have to be greater than zero).

In details, the following subsections introduce and discuss the simultaneous (AR) and conditional (CAR) *autoregressive* process as well as the *moving average* process (MA). Furthermore, in an excursion the *spatial lag* model and the *common factor constraint* are presented. The spatial lag model is a special case of the simultaneous spatial autoregressive process without a common factor constraint. However, this special case does not have a closed form of the co-variance matrix.

The following discussions will be conducted in terms of the placeholder spatial link matrix $\mathbf{V}$ which is allowed to be non-symmetric. Thus track needs to be kept on the transposition of the spatial link matrix. Attention will be laid on the underlying assumptions constituting these processes, their interrelationship to each other, the structure of their specific covariance matrices, and finally the feasible ranges of their autocorrelation parameter ρ. Use of this knowledge of the process properties, in particular the structure of the co-variance matrices and the feasible parameter range, helps later on to calculate the distribution of Moran's I under alternative hypotheses.

4.2.1   The Autoregressive Spatial Process

In brief, the simultaneous model is expressed by the joint probability density function between all observations given in equation (4.4) (see Anselin, 1988, pp 32–33). In contrast, a conditional autoregressive spatial model is based on a specification of the distribution of the response variable which is given in equation 4.1.

The Simultaneous Autoregressive Process The most widely used process in geographical analysis is the simultaneous AR-process. The correlogram of an AR-process fades out slowly whereas its partial correlogram approaches zero almost immediately. Such a pattern in the correlogram and partial correlogram is characteristic for an AR-process and can be used to identify an autoregressive spatial process (Cliff and Ord, 1981, p 139, Arbia, Chap. 8, 1989).

The formal definition of an autoregressive spatial process is given by

$$\mathbf{y} = \mathbf{X}\beta + \underbrace{\rho \cdot \mathbf{V} \cdot \varepsilon + \eta}_{=\varepsilon} \tag{4.7}$$

---

[3] Heteroscedasticity can easily be accommodated by adjusting the dependent variable $\mathbf{y}$ and independent variables $\mathbf{X}$ appropriately. This is done by a weighted least squares estimator. Then, however, the expectation of the regression residuals is no longer zero and therefore an additional intercept should be included in the model (see Kennedy, 1985, p 105).

$$= \mathbf{X}\boldsymbol{\beta} + \rho \cdot \mathbf{V} \cdot (\mathbf{y} - \mathbf{X}\boldsymbol{\beta}) + \boldsymbol{\eta} \text{ because } \varepsilon = \mathbf{y} - \mathbf{X}\boldsymbol{\beta} \qquad (4.8)$$

$$= \mathbf{X}\boldsymbol{\beta} + \rho \cdot \mathbf{V}\mathbf{y} - \mathbf{V}\mathbf{X} \cdot \rho \cdot \boldsymbol{\beta} + \boldsymbol{\eta} . \qquad (4.9)$$

The correlated error term ε, which is the only one which is indirectly observable via the regression residuals $\widehat{\varepsilon}$, can be defined by

$$\varepsilon \equiv (\mathbf{I} - \rho \cdot \mathbf{V})^{-1}\boldsymbol{\eta} . \qquad (4.10)$$

In terms of statistically independent disturbances $\boldsymbol{\eta} \sim N(\mathbf{0}, \sigma^2 \cdot \mathbf{I})$. Equation (4.9) illustrates that the dependent variable $\mathbf{y}$ in an autoregressive spatial model responds beside the disturbances $\boldsymbol{\eta}$ to a several other influences (Bailey and Gatrell, 1995, p 284). As is usual regression analysis, there are (i) the spatially independent influences of the exogenous component $\mathbf{X} \cdot \boldsymbol{\beta}$, then there are (ii) the spatially dependent endogenous observations $\rho \cdot \mathbf{V} \cdot \mathbf{y}$, finally there are (iii) the spatial trend values $-\mathbf{V} \cdot \mathbf{X} \cdot (\rho \cdot \boldsymbol{\beta})$. Note that the components of the coefficient $\rho \cdot \boldsymbol{\beta}$ are not directly identifiable.

Expression (ii) on the right hand side of the equation is critical. It depends via its expectation on the exogenous variables and via the random error term ε on the disturbances $\boldsymbol{\eta}$. It is therefore a random variable which is correlated with the disturbances. In fact, the co-variance between the endogenous variable and the disturbances is $\mathrm{Cov}(\mathbf{y}, \boldsymbol{\eta}) = \sigma^2 (\mathbf{I} - \mathbf{V}^T)^{-1}$ (see Cressie, 1991, p 406, or Haining, 1990, p 81). As Cliff and Ord (1981, appendix to Chap. 6) and Anselin (1988, p 58) have shown, the OLS estimator for the autocorrelation coefficient ρ is biased and inconsistent.

Equation (4.10) leads to an alternative formulation of an AR-process in terms of the independent disturbances $\boldsymbol{\eta}$ given by

$$\mathbf{y} = \mathbf{X}\boldsymbol{\beta} + (\mathbf{I} - \rho \cdot \mathbf{V})^{-1} \cdot \boldsymbol{\eta} \qquad (4.11)$$

$$(\mathbf{I} - \rho \cdot \mathbf{V}) \cdot \mathbf{y} = (\mathbf{I} - \rho \cdot \mathbf{V}) \cdot \mathbf{X}\boldsymbol{\beta} + \boldsymbol{\eta} . \qquad (4.12)$$

Note that equations (4.12) and (4.9) are equivalent. The variance-covariance matrix $\boldsymbol{\Omega}(\rho)$ between the error terms can be written as

$$\boldsymbol{\Omega}(\rho) = \mathrm{E}(\varepsilon \cdot \varepsilon^T) \qquad (4.13)$$

$$= \mathrm{E}\left[(\mathbf{I} - \rho \cdot \mathbf{V})^{-1}\boldsymbol{\eta} \cdot \boldsymbol{\eta}^T \left((\mathbf{I} - \rho \cdot \mathbf{V})^{-1}\right)^T\right] \qquad (4.14)$$

$$= (\mathbf{I} - \rho \cdot \mathbf{V})^{-1} \, \mathrm{E}(\boldsymbol{\eta} \cdot \boldsymbol{\eta}^T)\left((\mathbf{I} - \rho \cdot \mathbf{V})^{-1}\right)^T \qquad (4.15)$$

$$= (\mathbf{I} - \rho \cdot \mathbf{V})^{-1} \cdot \sigma^2 \cdot \mathbf{I} \cdot \left((\mathbf{I} - \rho \cdot \mathbf{V})^{-1}\right)^T \qquad (4.16)$$

$$= \sigma^2 \cdot (\mathbf{I} - \rho \cdot \mathbf{V})^{-1} \cdot \left((\mathbf{I} - \rho \cdot \mathbf{V})^T\right)^{-1} \qquad (4.17)$$

$$= \sigma^2 \cdot \left[(\mathbf{I} - \rho \cdot \mathbf{V})^T \cdot (\mathbf{I} - \rho \cdot \mathbf{V})\right]^{-1} \qquad (4.18)$$

$$\text{because } (\mathbf{A} \cdot \mathbf{B})^{-1} = \mathbf{B}^{-1}\mathbf{A}^{-1}$$

Clearly some restrictions are necessary to guarantee that $\boldsymbol{\Omega}(\rho)$ is positive definite. If the autocorrelation parameter ρ is within its feasible range

$$\rho \in \left] \frac{1}{\lambda_{\min}}, \frac{1}{\lambda_{\max}} \right[\qquad (4.19)$$

where $\lambda_{\min}$ and $\lambda_{\max}$ are the smallest and largest eigenvalues of $\mathbf{V}$, respectively, positive definiteness of the co-variance matrix of an AR-process is assured (see Anselin, 1982, Kelejian and Robinsion, 1995, or Hepple, 1995b). As long as the autocorrelation parameter ρ is within its feasible range the co-variance matrix of a simultaneous AR-process can be expanded in a converging Laurent series such that

$$(\mathbf{I} - \rho \cdot \mathbf{V})^{-1} \cdot (\mathbf{I} - \rho \cdot \mathbf{V}^T)^{-1} = \left(\mathbf{I} - \rho \cdot [(\mathbf{V}^T + \mathbf{V}) - \rho \cdot \mathbf{V}^T \cdot \mathbf{V}]\right)^{-1} \qquad (4.20)$$

$$= \mathbf{I} + \sum_{k=1}^{\infty} \rho^k \cdot [(\mathbf{V}^T + \mathbf{V}) - \rho \cdot \mathbf{V}^T \cdot \mathbf{V}]^k . \quad (4.21)$$

Without loss of generality it has been assumed that $\sigma^2 = 1$. Equation (4.21) shows that due to k-th power, higher order links partially determine the simultaneous AR-process. However, their impact declines because the weight ρ^k in the infinite series shrinks (see Haining, 1979, or Bivand, 1984).

A maximum likelihood estimator for the autoregressive spatial process comes with the advantage that the explicit specification of the co-variance matrix in the exponent of equation (4.5) does not need to be inverted because $\mathbf{\Omega}^{-1}(\rho) = \sigma^2 \cdot (\mathbf{I} - \rho \cdot \mathbf{V})^T \cdot (\mathbf{I} - \rho \cdot \mathbf{V})$. The evaluation of the Jacobian term $|\mathbf{\Omega}(\rho)|^{-\frac{1}{2}}$ can be simplified by using the spectrum of eigenvalues $\{\lambda_1, \ldots, \lambda_n\}$ of the spatial link matrix $\mathbf{V}$. Ord (1975) was the first to notice that $|\mathbf{I} - \rho \cdot \mathbf{V}| = \prod_{i=1}^{n}(1 - \rho \cdot \lambda_i)$. This expression shows also that Gaussian spatial processes are not defined at $\rho = \frac{1}{\lambda_i}$ for all $\lambda_i \in \{\lambda_1, \ldots, \lambda_n\}$. Note that only under specific circumstances, the spectrum of eigenvalues of a non-symmetric spatial link matrix does not involve imaginary numbers.

The Conditional Autoregressive Process The co-variance matrix of a conditional AR-process is defined by

$$\mathbf{\Omega}(\rho) = \mathrm{E}(\varepsilon \cdot \varepsilon^T) \qquad (4.22)$$

$$= \sigma^2 \cdot (\mathbf{I} - \rho \cdot \mathbf{V})^{-1} . \qquad (4.23)$$

It is obvious that the structure of this co-variance matrix is quite restrictive: the coded spatial link matrix $\mathbf{V}$ has to be symmetric to ensure positive definiteness of the co-variance matrix. For this reason its feasible range is restricted to $\rho \in \left] \frac{1}{\lambda_{\min}}, \frac{1}{\lambda_{\max}} \right[$. Due to this restriction the conditional AR-process is virtually not used in regional science, geography and spatial statistics despite the fact that the disturbances and the endogenous variables are independent, i.e., $\mathrm{Cov}(\mathbf{y}, \eta) = 0$ (see Cliff and Ord, 1981, appendix to Chap. 6). Furthermore, Cliff and Ord (1981, appendix to Chap. 6) have given a detailed derivation on how to obtain from the joint distribution the conditional distributions of the single observations and vice versa. They also have shown (1981, p 148) how the spatial link matrix of a conditional AR-process is related to a link matrix of a simultaneous AR-process which involves higher order spatial links.

In time series analysis conditional and simultaneous specified autoregressive processes result in an equivalent model due to the uni-directional arrangements of the consecutive time points. The literature does not agree whether the conditional or the simultaneous model is more appropriate to model hypothetical spatial relations. Cressie (1991, p 410) argues that spatial econometricans estimate their models in the simultaneous specification while they interpret the estimates from the conditional perspective.[4] If they do so, then they should better adopt a conditional specification in the first place. In contrast, Anselin (1988, p 33) argues that "most modeling situations in empirical regional science are easier expressed in a simultaneous form. This form is also closer to the standard econometric approach". Since in time-series econometrics the conditional and simultaneous autoregressive models are equivalent, Anselin's statement is questionable and spatial statistics and econometrics should be open to all kinds of spatial processes.

4.2.2   The Moving Average Process

The basic model structure of a moving average spatial process in the disturbances of a regression model is given by

$$\mathbf{y} = \mathbf{X}\boldsymbol{\beta} + \underbrace{\rho \cdot \mathbf{V} \cdot \boldsymbol{\eta} + \boldsymbol{\eta}}_{=\varepsilon} . \tag{4.24}$$

The error vector is defined in terms of the spatial process and the disturbances by

$$\varepsilon \equiv (\mathbf{I} + \rho \cdot \mathbf{V}) \cdot \boldsymbol{\eta} \tag{4.25}$$

where $\eta \sim N(0, \sigma^2 \cdot \mathbf{I})$. The co-variance matrix between the error terms can be written as

$$\boldsymbol{\Omega}(\rho) = \mathrm{E}(\varepsilon \cdot \varepsilon^T) \tag{4.26}$$

$$= \mathrm{E}[(\mathbf{I} + \rho \cdot \mathbf{V})\boldsymbol{\eta} \cdot \boldsymbol{\eta}^T (\mathbf{I} + \rho \cdot \mathbf{V})^T] \tag{4.27}$$

$$= \sigma^2 \cdot (\mathbf{I} + \rho \cdot \mathbf{V}) \cdot (\mathbf{I} + \rho \cdot \mathbf{V})^T . \tag{4.28}$$

The co-variance between the endogenous variable and the disturbances is $\mathrm{Cov}(\mathbf{y}, \eta) = \sigma^2(\mathbf{I} + \mathbf{V}^T)$ (see Haining, 1990, p 83). Thus the same comments related to the simultaneous autoregressive process apply for the moving average process. The feasible range of the moving average autocorrelation parameter ρ is

$$\rho \in \left] -\frac{1}{\lambda_{\max}}, -\frac{1}{\lambda_{\min}} \right[\tag{4.29}$$

---

[4] In the simultaneous model an interpretation of the spatial effects must focus solely on the structure of the error terms ε. Contrary, in the conditional model the expectation of the response variable has to be read in a conditional framework (i.e., $\mathrm{E}[Y_i \mid y_1, \ldots, y_{i-1}, y_{i+1}, \ldots, y_n]$ where $y_1, \ldots, y_{i-1}, y_{i+1}, \ldots, y_n$ are the observed values at locations other than i).

where $\lambda_{\min}$ and $\lambda_{\max}$ are the smallest and largest eigenvalues of $\mathbf{V}$, respectively. In contrast to the simultaneous AR-process, the MA-process comes with the disadvantage that the inverse co-variance matrix $\mathbf{\Omega}^{-1}(\rho)$ in the exponent of equation (4.5) needs to be evaluated explicitly.

In Anselin's opinion (1988, p 34) it is not worth following up the investigation of MA-processes because they do not differ substantially from AR-processes. Furthermore, he feels that the autoregressive form is simpler. To the author's knowledge there are only few systematic investigations in spatial statistics on the properties of a spatial MA-process. Notably Haining's work (1978) has to be mentioned here, who investigated the invertible and stationarity of a spatial moving average process. Tiefelsdorf, Griffith and Boots (1999) have shown that both processes differ substantially with respect to the tessellation induced heterogeneity. Hepple (1995a and 1995b) has devised a Bayesian test to distinguish between moving average and autoregressive processes. He shows that an underlying data generating process may very well have more in common with a moving average process.

4.2.3   Excursion: The Common Factor Constraint and the Spatial Lag Model

The spatial lag model can be conceived as a special case of the simultaneous AR-process. The simultaneous AR-process requires that the estimated coefficient $\widehat{\rho \cdot \beta}$ of expression $\mathbf{V} \cdot \mathbf{X} \cdot (-\rho \cdot \beta)$ satisfies the constraint $\widehat{\rho \cdot \beta} = \widehat{\rho} \cdot \widehat{\beta}$ where $\widehat{\beta}$ is the estimate of the expression $\mathbf{X} \cdot \beta$ and $\widehat{\rho}$ is the estimate of expression $\rho \cdot \mathbf{V} \cdot \mathbf{y}$. This *common factor constraint* has been extensively discussed in the spatial econometrics literature (see Burridge, 1981, Bivand, 1984, and Anselin, 1988). If this restriction is relaxed, equation (4.9) can be written as

$$\mathbf{y} = \mathbf{X} \cdot \beta + \rho \cdot \mathbf{V} \cdot \mathbf{y} + \mathbf{V} \cdot \mathbf{X} \cdot \gamma + \eta \qquad (4.30)$$

where $\gamma \neq \widehat{\rho} \cdot \widehat{\beta}$.

If on the other hand the constraint $\gamma = 0$ is imposed the simultaneous AR-process reduces to *spatial lag model*

$$\mathbf{y} = \mathbf{X}\beta + \rho \cdot \mathbf{V} \cdot \mathbf{y} + \eta \qquad (4.31)$$

where the spatial dependency is totally attributable to the interrelationship between the endogenous variable $\mathbf{y}$. A problem with both models is that they can not be specified in terms of a closed covariance matrix $\mathbf{\Omega}(\rho)$. This does not prohibit their estimation in a maximum likelihood framework but it proscribes the formulation of exact test methods for spatial autocorrelation in a ratio of quadratic forms.

4.3   Testing for Gaussian Spatial Processes

This section embeds the Moran's I test statistic into the general theory of statistical tests. It will outline that under the assumption of spatial independence

Moran's I is a *locally best invariant test* (King, 1981) and asymptotically a *Lagrangian Multiplier test* (Burridge, 1980) against a hypothetical spatial Gaussian process in either the MA- and AR-specification. King (1985a) provides a discussion in general terms on locally best invariant tests (LBI), the Lagrangian Multiplier (LM) test principle, and on two-sided hypothesis for arbitrary autocorrelated processes. In contrast, this discussion is restricted on one-sided parametric hypotheses about the spatial autocorrelation level ρ, i.e., either $H_0 : \rho_0 \leq 0$ against $H_1 : \rho > 0$ for positive spatial autocorrelation; or $H_0 : \rho_0 \geq 0$ against $H_1 : \rho < 0$ for negative spatial autocorrelation.

In a leap ahead, the spatial autocorrelation test statistic Moran's I needs to be formally introduced here. Part II will discuss its properties thoroughly in statistical terms. Moran's I is defined as scale invariant ratio of quadratic forms (RQF) in the normal distributed regression residuals $\widehat{\varepsilon}$, i.e.,

$$I = \frac{\widehat{\varepsilon}^T \cdot \frac{1}{2}(\mathbf{V} + \mathbf{V}^T) \cdot \widehat{\varepsilon}}{\widehat{\varepsilon}^T \cdot \widehat{\varepsilon}} . \tag{4.32}$$

Its expectation and variance under the assumption of spatial independence are given by

$$\mathrm{E}[I] = \frac{\mathrm{tr}(\mathbf{M} \cdot \mathbf{V})}{n - k} \text{ and} \tag{4.33}$$

$$\mathrm{Var}[I] = \frac{\mathrm{tr}(\mathbf{M} \cdot \mathbf{V} \cdot \mathbf{M} \cdot \mathbf{V}^T) + \mathrm{tr}(\mathbf{M} \cdot \mathbf{V})^2 + \{\mathrm{tr}(\mathbf{M} \cdot \mathbf{V})\}^2}{(n - k) \cdot (n - k + 2)} - \{\mathrm{E}[I]\}^2 , \tag{4.34}$$

respectively, where $\mathrm{tr}(\cdot)$ denotes the trace operator, $\mathbf{M}$ the projection matrix and $\mathbf{V}$ the spatial link matrix. With these parameters the z-transformed Moran's I, i.e.,

$$z(I) = \frac{I - \mathrm{E}[I]}{\sqrt{\mathrm{Var}[I]}} , \tag{4.35}$$

can be calculated. This statistic is for normal distributed regression residuals and well-behaved spatial link matrices under the assumption of spatial independence asymptotically normal distributed with $z(I) \underset{\mathrm{asympt.}}{\sim} \mathcal{N}(0, 1)$. A value of $z(I)$ greater than zero indicates positive spatial autocorrelation where residuals of equal sign tend to cluster spatially. A value of $z(I)$ below zero points to a spatial pattern in the regression residuals where their positive and negative signs alternate spatially from one location to the other.

First the LM-test principle for spatial autocorrelation is introduced and compared against the likelihood ratio (LR) test principle. Both test principles are based on an asymptotic concept derived from the likelihood function and their small sample properties are unknown. With these results some statistical quality properties of test statistics like 'uniform most powerful' and 'locally best invariant' can be addressed. A general review of the statistical test theory can be found, for instance, in Mood, Graybill and Boes (1974) and a didactical introduction with a discussion on the LR- and LM-test principle is given in Buse (1982).

4.3.1   Test Principles for Spatial Autocorrelation in Regression Residuals

Recall that the linear model with spatially autocorrelated and normal distributed disturbances is written as $\mathbf{y} = \mathbf{X} \cdot \boldsymbol{\beta} + \varepsilon$ with $\varepsilon \sim \mathcal{N}(0, \boldsymbol{\Omega}(\rho))$. Under the alternative hypothesis the likelihood function $f(\mathbf{y} \mid \rho, \boldsymbol{\beta}, \sigma^2)$ of the realization $\mathbf{y}$ with respect to the parameters ρ, $\boldsymbol{\beta}$, σ^2 and the regression matrix $\mathbf{X}$ is expressed by

$$f(\mathbf{y} \mid \rho, \boldsymbol{\beta}, \sigma^2) = (2\pi)^{-\frac{n}{2}} \cdot |\boldsymbol{\Omega}(\rho)|^{-\frac{1}{2}} \cdot \exp\left(-\frac{1}{2}(\mathbf{y} - \mathbf{X}\boldsymbol{\beta})^T \cdot \boldsymbol{\Omega}^{-1}(\rho) \cdot (\mathbf{y} - \mathbf{X}\boldsymbol{\beta})\right)$$

(4.36)

for the n observations. Taking the natural logarithm of the likelihood function the log-likelihood function

$$\mathcal{L}(\rho, \boldsymbol{\beta}, \sigma^2) \equiv \ln f(\mathbf{y} \mid \rho, \boldsymbol{\beta}, \sigma^2)$$

(4.37)

can be calculated which is algebraically easier to handle. Based on the log-likelihood function in equation (4.37) the definition of the LR-test for $H_0 : \rho_0 = 0$ is given by

$$\lambda_{(\rho)}^{LR} \equiv -2 \cdot \left(\mathcal{L}(0, \boldsymbol{\beta}, \sigma^2) - \sup_\rho \mathcal{L}(\rho, \boldsymbol{\beta}, \sigma^2)\right).$$

(4.38)

$\mathcal{L}(0, \boldsymbol{\beta}, \sigma^2)$ symbolizes that the log-likelihood function is evaluated under the restriction $\rho_0 = 0$ which is equivalent to the assumption of spatial independence, whereas $\sup_\rho \mathcal{L}(\rho, \boldsymbol{\beta}, \sigma^2)$ denotes the value of the log-likelihood function maximized with respect to ρ (i.e., the estimated value is given by $\widehat{\rho} = \rho_{\max}$). Both log-likelihood functions are already maximized with respect to the parameters $\boldsymbol{\beta}, \sigma^2$ and the relation $\sup_\rho \mathcal{L}(\rho, \boldsymbol{\beta}, \sigma^2) \geq \mathcal{L}(0, \boldsymbol{\beta}, \sigma^2) \geq 0$ holds. If $\lambda_{(\rho)}^{LR}$ is significantly different from 0 then the restriction $\rho_0 = 0$ is vital and we have to reject the zero hypothesis. Asymptotically the LR-statistic $\lambda_{(\rho)}^{LR}$ is χ^2-distributed with one degree of freedom.

The slope of $\sup_\rho \mathcal{L}(\rho, \boldsymbol{\beta}, \sigma^2)$ (i.e., its first derivative with respect to the autocorrelation level ρ at $\widehat{\rho} = \rho_{\max}$) is zero. The LM-statistic is based on this slope. It is formulated as a restricted maximization of the augmented log-likelihood function (4.37) and is written as

$$\mathcal{L}_R(\rho, \boldsymbol{\beta}, \sigma^2, \lambda) \equiv \mathcal{L}(\rho_0, \boldsymbol{\beta}, \sigma^2) - \lambda \cdot (\rho_0 - 0),$$

(4.39)

where λ is the scalar Lagrangian Multiplier indicating the slope at the constraint $\rho_0 = 0$. The first order conditions for a stationary value of the restricted likelihood function $\mathcal{L}_R(\rho, \boldsymbol{\beta}, \sigma^2, \lambda)$ yield the familiar OLS estimates, $\widehat{\boldsymbol{\beta}}$ and $\widehat{\sigma}^2$, for $\boldsymbol{\beta}$ and σ^2, and an estimated value $\widehat{\lambda}$ at the constrained maximum. Due to the restriction the first derivative with respect to ρ of $\partial \mathcal{L}_R(\rho, \boldsymbol{\beta}, \sigma^2, \lambda)/\partial\rho$ equals zero automatically. If the estimated Lagrangian Multiplier $\widehat{\lambda}$ differs significantly from zero this indicates that the zero hypothesis is not valid.

Burridge (1980) has shown that the Lagrangian Multiplier $\widehat{\lambda}$ for the co-variance matrices $\Omega(\rho)$ of plain spatial Gaussian MA- and AR-processes is

$$\widehat{\lambda} = \frac{\widehat{\varepsilon}^T \cdot \frac{1}{2} \cdot (\mathbf{V} + \mathbf{V}^T) \cdot \widehat{\varepsilon}}{\widehat{\sigma}^2} \tag{4.40}$$

which is proportional to the Moran's I test statistic. Therefore, Moran's I is similar to a LM-test. To account for the sample variability we have to weight $\widehat{\lambda}$ by the curvature of the log-likelihood function at $\rho_0 = 0$. This adjustment factor is expressed by the information matrix

$$-E \left(\frac{\partial^2 \mathcal{L}(\rho, \beta, \sigma^2)}{\partial \rho \cdot \partial \rho} \right)\bigg|_{\rho_0=0} = \mathrm{tr}(\mathbf{V}^2 + \mathbf{V}^T \cdot \mathbf{V}) \tag{4.41}$$

evaluated at $\rho_0 = 0$. Notice that under the zero hypothesis the information matrix is block diagonal so that the cross derivatives $\partial \rho \cdot \partial \beta$ and $\partial \rho \cdot \partial \sigma^2$ can be ignored. However, as is common in the maximum likelihood framework, the impact of the projection matrix $\mathbf{M}$ onto the variance term $\mathrm{tr}(\mathbf{V}^2 + \mathbf{V}^T \cdot \mathbf{V})$ has been neglected. The standardized Lagrangian Multiplier statistic

$$\widehat{\lambda}^{LM} \equiv \frac{\widehat{\lambda}^2}{\mathrm{tr}(\mathbf{V}^2 + \mathbf{V}^T \cdot \mathbf{V})} \tag{4.42}$$

is again asymptotically χ^2-distributed with 1 degree of freedom and is indepen-dent of ρ. Besides the suppression of the projection matrix $\mathbf{M}$ the λ^{LM}-statistic also discards the non-zero expectation of Moran's I. It is my feeling that com-putational efforts to calculate the z-standardized Moran's I statistic is not as complex as claimed by Anselin and Florax (1995, p 21), especially when it is compared to the λ^{LM}-statistic which they favor.

Silvey (1959) has shown that a LM-test and a LR-test are asymptotic equiva-lent. Nevertheless, they satisfy the inequality $\lambda^{LM} \leq \lambda^{LR}$ which means that the LM-test is more conservative than the LR-test. The LM-test comes with the ad-vantage that the restricted log-likelihood function must be evaluated only once under the zero hypothesis of spatial independence which leads to the well-known OLS estimators for β and σ^2. In contrast, for the LR-test the parameters have to be estimated twice, once under the un-restricted fully specified model and once under the restriction.

Anselin (1988) strongly advocates the LM-test and has made extensive use of this principle to develop tests for many other specifications of spatial autore-gressive processes like the common factor constraint and the spatial lag model. Furthermore, Kennedy (1985, p 65) states that

"Because it requires estimation under only the null hypothesis, the LM-test is less specific than other tests concerning the precise nature of the alternative hypothesis. This could be viewed as an advantage, since it allows testing to be conducted in the context of a more general alterna-tive hypothesis, or as a disadvantage, since it does not permit the precise nature of an alternative hypothesis to play a role and thereby increase the power of the test. Monte Carlo studies, for example, have shown that this potential drawback is not of great concern."

Anselin and Rey (1991) and Anselin and Florax (1995) compared Moran's I by means of simulation experiments to LM-tests related to two specific alternative hypotheses. Their experiments (Anselin and Florax, 1995, p 22) "also indicated a tendency for this test (i.e., Moran's I) to have power against several types of alternatives, including non-normality, heteroscedasticity and different forms of spatial dependence". Their first model was based on a plain autoregressive (or moving average) spatial process and they came to the conclusion that Moran's I has in general a higher power than the λ^{LM}-statistic in equation (4.42) for AR- and MA-processes. This finding comes as no surprise because the definition of the z-standardized Moran's I also utilizes the information in the projection matrix $\mathbf{M}$. Their second model was specified as spatial autoregressive lag model (i.e., $(\mathbf{I} - \rho \cdot \mathbf{V}) \cdot \mathbf{y} = \mathbf{X} \cdot \boldsymbol{\beta} + \eta$ with $\eta \sim \mathcal{N}(\mathbf{0}, \sigma^2 \cdot \mathbf{I})$) and they used a LM-test specific for this model (see Anselin, 1988, for a derivation of this test). As expected, here the power of this specific LM-test outperforms that of Moran's I.

In other simulation studies Cliff and Ord (1975) and Haining (1977, 1978) found that the likelihood ratio test λ^{LR} is more powerful than Moran's I, although the difference is not marked until ρ is relatively large. However, as outlined below, for AR- and MA-processes Moran's I is in the neighborhood of the zero hypothesis most powerful. In conclusion, Moran's I is a *good general purpose* test for spatial autocorrelation in AR- and MA-processes in particular and model violations and other spatial processes in general. Nonetheless, for descendant specifications of these intrinsic processes there are more specialized tests available.

4.3.2   Statistical Properties of the Moran's I Test

The quality of statistical tests is usually evaluated by means of their *power function* under pre-specified alternative hypotheses. Power functions measure the probability of rejecting the zero hypothesis for a given significance level (in theoretical statistics also called size of a test) when in fact the distribution from where the sample has been drawn is specified by a parameter not belonging to the zero hypothesis. In the case that the parameter belongs to the zero hypothesis the power function shall always be smaller than the given significance level. Whereas, in case that the alternative hypothesis is true, as the parameter under consideration deviates more and more from the zero hypothesis the more extreme the power function shall become and finally approach 1. The faster the power function approaches 1 the better. Clearly, a test whose power function under the alternative hypothesis is always larger than those of competing tests is preferred because it has a higher potential to reject the zero hypothesis when in fact it is false. Such a test is called *uniformly most powerful* (UMP). The term 'uniformly' refers here to the whole range of the parameter under the alternative hypothesis. For an alternative use of the power function see Chapter 11.3.1.

Unfortunately, there is no such test for spatial autocorrelation which is UMP. So we have to search for a test that preserves at least some of the properties of a UMP test. Since Moran's I belongs to the class of scale *invariant* statistics it can be shown that it is at least a *locally best invariant* test (LBI). Only under very specific conditions is Moran's I also a UMP invariant test. The following discussion will heuristically delineate why Moran's I can not be a UMP test. Then it outlines that the one-sided Moran's I test is at least in either the positive or negative neighborhood of $\rho_0 = 0$, where it matters most, most powerful in the sense that its power function has maximum slope there.

Uniformly Most Powerful Tests The general aim of any test strategy is to separate the sample space $\mathfrak{Y}$, i.e., the values a random vector Y under consideration can take, into two subsets. Subset $\mathfrak{Y}_0$ is supporting the zero hypothesis and subset $\mathfrak{Y}_1$ is associated with the alternative hypothesis and called the *critical region*. Usually this strategy is comprised into a sufficient test statistic Υ which partitions $\mathfrak{Y}$ into the two unknown subsets. Let us consider first the simple hypothesis that the random vector Y comes either from the distribution $f(y \mid \omega_0)$ or the distribution $f(y \mid \omega_1)$. Then the Neyman-Pearson lemma helps to design a test statistic Υ with optimal properties.

Lemma 3 (Neyman-Pearson). *Let α be the size of the test. Then the probability of erroneously rejecting the zero hypothesis is*

$$\Pr(y \in \mathfrak{Y}_1 \mid \omega_0) = \alpha \ . \tag{4.43}$$

Furthermore, let the ratio of the likelihood functions under the zero and the alternative hypothesis be

$$\lambda = \frac{f(y \mid \omega_0)}{f(y \mid \omega_1)} \lessgtr k \tag{4.44}$$

such that $\lambda \leq k$ if $Y \in \mathfrak{Y}_1$ and $\lambda > k$ if $Y \in \mathfrak{Y}_0$. Then, the test statistic Υ corresponding to the critical region $\mathfrak{Y}_1$ is a most powerful test of size α of $H_0 : \omega = \omega_0$ versus $H_1 : \omega = \omega_1$.

Notice that equation (4.43) still permits different subsets $\mathfrak{Y}_1$ satisfying the size α of the test. Only condition (4.44) imposes the necessary constraint to construct a test Υ discriminating between the critical region $\mathfrak{Y}_1$ and $\mathfrak{Y}_0$. The test statistic Υ is based on the size α and is a function of the random vector Y as well as the distribution function $f(y \mid \omega)$. However, the distribution of Υ still needs to be derived and, as in the case of Moran's I, this is not done easily. If the test statistic Υ associated with the critical region $\mathfrak{Y}_1$ does not depend on ω_1 for any $\omega_1 \neq \omega_0$ then it is said to be *uniformly most powerful*. Notice that the likelihood ratio test (4.38) is closely related to the ratio of the likelihood functions in (4.44). Nevertheless, the likelihood ratio test depends also on ω_1 and its small sample distribution is unknown.

The Locally Best Invariant Statistic Moran's I Unfortunately, it is not possible (with an exception) to design a test statistic Υ of spatial autocorrelation which is based on the likelihood function in equation (4.36) and which is independent from the autocorrelation level ρ (see King, 1985). Earlier Anderson (1948) showed that only if the underlying density function under the alternative hypothesis is of form

$$c \cdot \exp\left[-\frac{1}{2\cdot\sigma^2}\{(1+\rho^2)\varepsilon^T\cdot\varepsilon - 2\rho\cdot\varepsilon^T\cdot\mathbf{\Theta}\cdot\varepsilon\}\right] \tag{4.45}$$

and the regression matrix $\mathbf{X}$ consists solely of latent vectors of the symmetric structure matrix $\mathbf{\Theta}$ then the test statistic

$$\Upsilon_r \equiv \frac{\widehat{\varepsilon}^T\cdot\mathbf{\Theta}\cdot\widehat{\varepsilon}}{\widehat{\varepsilon}^T\cdot\widehat{\varepsilon}} \tag{4.46}$$

is UMP. The parameter c is a non-relevant constant. Durbin and Watson (1950) made use of this and approximated the distribution (4.45) to the error distribution of an autoregressive Markov process by an appropriate specification of the structure matrix $\mathbf{\Theta}$. Shifting to the spatial situation, also here the distribution function of a spatial process defined in the spatial co-variance matrix $\mathbf{\Omega}(\rho)$ can not be brought into the form (4.45). This illustrates that there is no algebraic *one-to-one* link between the Moran's I test statistic and the autocorrelation parameter of the MA- or AR-process.

To amend this unsatisfactory situation King (1981 and 1985a) restricted his attention to scale invariant tests and made use of results by Ferguson (1967, p 235). The test statistic Υ_s defined by

$$\Upsilon_s \equiv -\frac{1}{2}\frac{\widehat{\varepsilon}^T\cdot\left.\frac{\partial\mathbf{\Omega}^{-1}(\rho)}{\partial\rho}\right|_{\rho_0=0}\cdot\widehat{\varepsilon}}{\widehat{\varepsilon}^T\cdot\widehat{\varepsilon}} \tag{4.47}$$

is scale invariant under the transformation

$$\mathbf{y} \to \widetilde{\mathbf{y}} = \gamma_0\cdot\mathbf{y} + \mathbf{X}\cdot\widetilde{\gamma}, \tag{4.48}$$

with γ_0 being a positive scalar, $\widetilde{\gamma}$ a $p\times 1$ real vector and $\widehat{\widetilde{\varepsilon}} = \mathbf{M}\cdot\widetilde{\mathbf{y}}$. King has shown that scale invariant test statistics Υ_s are at least LBI-tests in the neighborhood of $\rho_0 = 0$ for processes having a distribution function (4.36). For plain MA as well as the AR spatial processes the derivatives matrix at $\rho_0 = 0$ is $\left.\frac{\partial\mathbf{\Omega}^{-1}(\rho)}{\partial\rho}\right|_{\rho_0=0} = -(\mathbf{V}+\mathbf{V}^T)$ because $\mathbf{\Omega}^{-1}(0) = \mathbf{I}$.

Consequently, Moran's I is a LBI-test under the zero hypothesis against a hypothetical spatial AR-process as well as against a hypothetical spatial MA-process. Also Moran's I's property of being a LBI-test is independent of the choice of the spatial link matrix and the employed coding scheme. The gradient of Moran's I power function is in the neighborhood of the zero hypothesis steeper than that for any other test. This border line between the sample space $\mathfrak{Y}_0$ under the zero hypothesis and the critical region $\mathfrak{Y}_1$ under the alternative hypothesis is most pivotal for statistical tests as long as the power function is well-behaved monotone. Nevertheless, at $\rho_0 \neq 0$ there will be other test statistics which are point optimal under specific alternative hypotheses and possess higher power there than Moran's I (recall the results of the simulation experiments by Cliff and Ord as well as Haining mentioned at the bottom of Section 4.3.1). King (1985b and 1985c) as well as Dufour and King (1991) have constructed two point optimal test statistics for serial autocorrelation (i.e., $\rho_0 \neq 0$, one specific for an MA-process and one specific for an AR-process).

King (1981) points out a special case for the Moran's I test statistic. If the structure matrix Θ is specified such that its latent vectors coincide with the regression matrix $\mathbf{X}$ then invariant tests of the form (4.46) are also UMP. One could speculate that the coded link matrix $\mathbf{W}$ would lead to a UMP test if the regression matrix $\mathbf{X}$ consists only of the unity vector $\mathbf{1}$. However, even though the first latent vector of a spatial link matrix in the W-coding is constant and therefore coincides with the unity vector $\mathbf{1}$ of a regression matrix, the argumentation must be based on the symmetrized form $(\mathbf{W} + \mathbf{W}^T)/2$ which does not possess a constant latent vector.

Part II

Development and Properties of Moran's I's Distribution

In order to derive a *closed theory* on the exact distribution of Moran's I either under the assumption of spatial independence or conditional upon a significant spatial process this Part has to be rather technical. It will start with results from mathematical statistics. These show the relation between a distribution function and its characteristic function. The characteristic function allows one to derive the distribution of a weighted sum of χ^2-distributed random variables as shown by Imhof. The distribution of Moran's I, which is a ratio of quadratic forms in normal distributed variables, can be brought into this form after a sequence of transformation. This has been first demonstrated by Koerts and Abrahamse. Before a statistical test can be conducted it is mandatory to know its exact distribution or at least some sensible approximation of it. However, often knowledge on the parameters of its distribution also proves to be useful. Here the bounds of Moran's I's value range and its moments shall given. These allow one to compare properties of Moran's I exact distribution under the independence assumption and under a spatial process. Extensions to the pair-wise correlation between Moran's Is defined on the same regression model and spatial process but using different spatial structures can be given. Other extensions, such as the multivariate distribution of several ratios of quadratic form, are summarized.

Furthermore, the evaluation of the exact distribution and moments involves numerical integration, thus some numerical aspects will be discussed. Because numerical integration is a resource demanding computational method, small and large sample properties are compared against each other and the performance of the normal approximation is investigated for well-behaved spatial situations. It should be stressed at this point that numerical methods become less and less prohibiting nowadays with constant increase in present computing technology and numerical algorithms. In fact, the statistical methods presented here are not new at all and are mostly known for over twenty years (see Figure 1.1), though only recently they became computationally feasible. The exact distribution of Moran's I can be calculated swiftly on present high-end personal computers for several hundred spatial objects. However, for massive exploratory studies, such as an analysis by means of local Moran's I_i, still some patience is required.

Chapter 5

The Characteristic Function

The characteristic function $\phi(t)$ is a generalization of the moment generating function $m(t)$ which is used to evaluate the moments of distribution functions. Since the moment generating function $m(t)$ does not always exists it has been lifted into the complex-valued plane where it always exists and is called there the characteristic function $\phi(t)$. The characteristic function $\phi(t)$ can not only be used to calculate the moments of a distribution function $F(x)$; since there is a one-to-one relationship between the distribution function and the characteristic function it also plays a significant role in deriving new distribution functions. Epps (1993) gives a good geometrical illustration of the characteristic function and mentions further applications.

In fact, the relation between the characteristic function and the distribution function will be used here to calculate the exact distribution of Moran's I, whose characteristic function is known but not its explicit distribution function. This section will proceed by defining the characteristic function and summarizing some of its properties. Then the inversion theorem is presented and proven. Finally, use of the inversion theorem is made. The distribution function of a weighted sum of independent central χ^2-distributed variables, whose characteristic function is known, will be derived and simplified to the well-known Imhof's formula.

5.1   General Properties of the Characteristic Function

The characteristic function $\phi(t) \equiv \mathrm{E}[e^{itx}]$ is defined for any real number t and random variable X, with the underlying distribution function $F(x)$, as the expectation of the complex valued transformation e^{itx} where i denotes the imaginary unit with $i^2 = -1$. This Fourier transformation $e^{itx} \equiv \cos(tx) + i \cdot \sin(tx)$ takes the realization of X from the real line to the perimeter of the unit circle in the complex plane. Therefore, following Epps (1993) the value $\phi(t)$ can be interpreted as the center of mass of the density function $f(x)$ wrapped around the unit circle in the complex plane at the value t.

In details the characteristic function $\phi(t)$ for a discrete-type or continuous-type random variable X is defined as

$$\phi(t) \equiv \begin{cases} \sum_{x_j \in \mathfrak{X}} e^{itx_j} \cdot p(x_j) & \text{for a discrete-type random variable} \\ \int_{x \in \mathfrak{X}} e^{itx} \cdot f(x) \cdot dx & \text{for a continuous-type random variable} \end{cases} \tag{5.1}$$

where $\mathfrak{X}$ stands for the sample space of the random variable, e.g., for the Bernoulli distribution $\mathfrak{X} = \{0, 1\}$ and for the normal distribution $\mathfrak{X} = [-\infty, \infty]$.

The basic properties of the characteristic function $\phi(t)$ are as follows:

1. $\phi(t)$ always exists because $|e^{itx}|$ is a continuous and bounded function for all finite real values of t and x.
2. $\phi(0) = 1$ for any distribution function.
3. $|\phi(t)| = |\mathrm{E}(e^{itx})| < 1$ for any non-degenerated distribution function and $t \neq 0$.
4. $\phi(-t) = \phi^*(t)$, where $\phi^*(t)$ denotes the complex conjugate to $\phi(t)$ (i.e., we only need to concentrate on $t \geq 0$ because $\phi(t)$ is symmetric around $t = 0$).

Without proofs the following theorems are given. Proofs and examples can for instance be found in Ochi (1990).

Theorem 3. *The r-th moment of a random variable, $\mathrm{E}[x^r]$, if it exists, can be obtained by*

$$\mathrm{E}[x^r] = \frac{1}{i^r} \phi^{(r)}(0) , \qquad (5.2)$$

where i is the imaginary unit, and $\phi^{(r)}(0) \equiv \frac{d^r}{dt^r} \phi(t)|_{t=0}$ is the r-th derivative of the characteristic function around $t = 0$.

This theorem is an extension of the moment generating function to the characteristic function. It has been included only for completeness and will not be used further.

Theorem 4. *Let $\phi_x(t)$ be the characteristic function of a random variable X and consider a new random variable $Y = \lambda \cdot X + \mu$, where λ and μ are real constants. Then, the characteristic function of the random variable Y is given by*

$$\phi_y(t) = e^{it\mu} \cdot \phi_x(\lambda \cdot t) . \qquad (5.3)$$

The translation coefficient μ shifts the distribution and consequently its expectation. The coefficient λ re-scales the distribution and operates as weight of random variable X which, nevertheless, does not need to be positive.

Theorem 5. *Let the random variables $X_1, X_2, \ldots, X_n$ be stochastically independent. Then the characteristic function of the new random variable $Y \equiv X_1 + X_2 + \cdots + X_n$ is the product of the characteristic functions $\phi_j(t)$ of each random variable. That is,*

$$\phi_y(t) = \prod_{j=1}^{n} \phi_j(t) . \qquad (5.4)$$

This theorem gives the characteristic function of a sum of independent random variables which, however, are allowed to have different distributions.

Corollary 1. *The combination of Theorems 4 and 5 allows one to calculate the characteristic function of n independent weighted random variables $Y = \lambda_1 \cdot X_1 + \lambda_2 \cdot X_2 + \cdots + \lambda_n \cdot X_n$ which is*

$$\phi_y(t) = \prod_{j=1}^{n} \phi_j(\lambda_j \cdot t) \tag{5.5}$$

with the weights λ_j being real coefficients.

Note that this calculation is done completely in the domain of characteristic functions so that the underlying distribution functions do not need to be known. Later use of this joint characteristic function will be made.

5.2   The Inversion Theorem

The inversion theorem links the characteristic function $\phi(t)$ back to its underlying distribution function $F(x)$. It comes in many different forms (see Stuart and Ord, 1994, p 125ff) and is stated below for continuous-type random variables in the form given by Gil-Peleaz (1951) which directly links it to the distribution function $F(x)$.

Theorem 6 (Gil-Peleaz). *Let $\phi(t)$ be the characteristic function of a random variable X. If X is a continuous-type random variable its distribution function can be evaluated by*

$$F(x) = \frac{1}{2} + \frac{1}{2\pi} \int_0^\infty \frac{e^{itx} \cdot \phi(-t) - e^{-itx} \cdot \phi(t)}{it} \cdot dt \ . \tag{5.6}$$

Corollary 2. *If two distribution functions share the same characteristic function for every $t \geq 0$ then they have to be identical.*

Therefore, the distribution function is uniquely 'characterized' by the characteristic function. Waller (1995) gives an example of this property of the characteristic function where two distinct characteristic functions distinguish two very similar appearing distribution functions.

Other relations stated in the literature are the initial form given by Lévy (1925)

$$F(x) - F(0) = \frac{1}{2\pi} \int_{-\infty}^\infty \frac{1 - e^{-itx}}{it} \cdot \phi(t) \cdot dt \ , \tag{5.7}$$

or in terms of the density function $f(x) = \frac{dF(x)}{dx}$

$$f(x) = \frac{1}{2\pi} \int_{-\infty}^\infty e^{-itx} \cdot \phi(t) \cdot dt \tag{5.8}$$

where the constant term $F(0)$ is eliminated by differentiation. This form illustrates the reciprocal relationship between $f(x)$ and $\phi(t) = \int_{-\infty}^\infty e^{itx} \cdot f(x) \cdot dx$.

Because the inversion formula is of fundamental importance for the rest of this discussion it will be proven here for the form of $F(x)$ (see equation 5.6). The proof below follows closely the one given by Koerts and Abrahamse (1969, p 77ff) which in turn rests on Gil-Peleaz (1951). In order to prove equation (5.6), let us consider the step-function

$$\text{sign}(y - x) = \frac{2}{\pi} \int_0^\infty \frac{\sin t(y - x)}{t} \cdot dt = \begin{cases} -1 & \text{if } y < x, \\ 0 & \text{if } y = x, \\ 1 & \text{if } y > x, \end{cases} \quad (5.9)$$

and observe that

$$\int_{-\infty}^\infty \text{sign}(y - x) \cdot f(y) \cdot dy = -\int_{-\infty}^x f(y) \cdot dy + \int_x^\infty f(y) \cdot dy \quad (5.10)$$
$$= -F(x) + \big(1 - F(x)\big)$$
$$= 1 - 2 \cdot F(x) .$$

Now, for any positive numbers a and b we have

$$\frac{1}{\pi} \int_a^b \frac{e^{itx} \cdot \phi(-t) - e^{-itx} \cdot \phi(t)}{it} \cdot dt$$

$$= \frac{1}{\pi} \int_a^b \int_{-\infty}^\infty \frac{e^{-it(y-x)} - e^{it(y-x)}}{it} \cdot f(y) \cdot dy \cdot dt \quad (5.11)$$

$$= -\frac{2}{\pi} \int_a^b \int_{-\infty}^\infty \frac{\sin t(y - x)}{t} \cdot f(y) \cdot dy \cdot dt \quad (5.12)$$

$$= -\int_{-\infty}^\infty f(y) \cdot \frac{2}{\pi} \int_a^b \frac{\sin t(y - x)}{t} \cdot dt \cdot dy \quad (5.13)$$

where in equation (5.11) the definition of the characteristic function $\phi(t) = \int_{-\infty}^\infty e^{ity} \cdot f(y) \cdot dy$ has been applied and in equation (5.12) use is made of $\frac{e^{-iv} - e^{iv}}{i} = -2\sin v$. The change in the order of integration in equation (5.13) is possible because for all t in $[a, b]$ and for all y it holds that $\frac{\sin t(y-x)}{t}$ is continuous with respect to t and that it is bound, i.e.,

$$\left| \frac{\sin t(y - x)}{t} \right| \leq \frac{|\sin t(y - x)|}{|t|} \leq \frac{1}{|t|} < \frac{1}{a} . \quad (5.14)$$

When a tends to zero and b tends to infinity, the integral

$$-\int_{-\infty}^\infty f(y) \cdot \lim_{\substack{a \to 0 \\ b \to \infty}} \frac{2}{\pi} \int_a^b \frac{\sin t(y - x)}{t} \cdot dt \cdot dy = 2F(x) - 1 \quad (5.15)$$

tends to equation (5.10) which proves the inversion theorem given in formula (5.6). We are allowed to take the limits under the integral sign because

$\int_a^b \frac{\sin t(y-x)}{t} \cdot dt$ is always finite and bound, i.e.,

$$\left| \int_a^b \frac{\sin t(y-x)}{t} \cdot dt \right| \leq \int_a^b \left| \frac{\sin t(y-x)}{t} \right| \cdot dt \leq \int_a^b \frac{1}{|t|} \leq \frac{b-a}{a} . \qquad (5.16)$$

Another useful simplification of $F(x) = \frac{1}{2} + \frac{1}{2\pi} \int_0^\infty \frac{e^{itx} \cdot \phi(-t) - e^{-itx} \cdot \phi(t)}{it} \cdot dt$ can be performed by making use of

$$\int_0^\infty \int_{-\infty}^\infty \frac{e^{-it(y-x)} - e^{it(y-x)}}{it} \cdot f(y) \cdot dy \cdot dt \qquad (5.17)$$

$$= -2 \int_0^\infty \int_{-\infty}^\infty \frac{\sin t(y-x)}{t} \cdot f(y) \cdot dy \cdot dt$$

and noting that

$$\frac{1}{t} \cdot e^{-itx} \cdot \phi(t) = \int_{-\infty}^\infty \frac{1}{t} \cdot e^{-it(y-x)} \cdot f(y) \cdot dy \qquad (5.18)$$

$$= \int_{-\infty}^\infty \frac{\cos t(y-x)}{t} \cdot f(y) \cdot dy - i \int_{-\infty}^\infty \frac{\sin t(y-x)}{t} \cdot f(y) \cdot dy$$

Thus the inversion formula (5.6) reduces to

$$F(x) = \frac{1}{2} - \frac{1}{\pi} \int_0^\infty \frac{1}{t} \Im\left(e^{-itx} \cdot \phi(t) \right) \cdot dt \qquad (5.19)$$

where the function $\Im(\cdot)$ stands for the imaginary part of a complex number.

5.3   Distribution of a Weighted Sum of χ^2-Distributed Variables

This section will concentrate solely on the derivation of the distribution of a weighted sum of *central* χ^2-distributed variables, since no use of non-central χ^2-distributed variables[1] is made. By doing so, we gain a substantial simplification in the derivation and calculation of the exact distribution function. A general derivation can be found in Imhof (1961), who has to be credited for developing the algebraic simplification of the inversion theorem for weighted sums of non-central χ^2-distributed variables. Koerts and Abrahamse (1969, p 82–85) give a more detailed derivation.

It is known that the square of a $\mathcal{N}(0,1)$-distributed random variable X is a central χ^2-distributed variable with one degree of freedom (see for instance Stuart and Ord, 1994, pp 374ff, for a derivation of the relationship). The characteristic function of a central χ^2-distributed variable X^2 is given by

---

[1] A non-central χ^2-distributed variable would arise if $X \sim \mathcal{N}(\mu, 1)$. Then the non-centrality parameter δ^2 is defined by $\delta^2 \equiv \mu^2$.

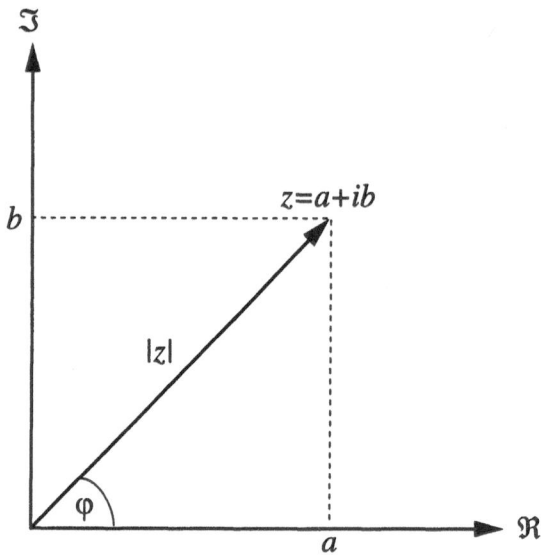

Fig. 5.1. Representation of complex numbers in the complex plain

$\phi_{x^2}(t) = (1 - 2it)^{-\frac{1}{2}}$. Therefore, using Theorems 4 and 5 the weighted sum of n central χ^2-distributed variables

$$Y \equiv \lambda_1 \cdot X_1^2 + \lambda_2 \cdot X_2^2 + \cdots + \lambda_n \cdot X_n^2 \qquad (5.20)$$

possesses the characteristic function

$$\phi_y(t) = \prod_{j=1}^{n} (1 - 2it\lambda_j)^{-\frac{1}{2}} . \qquad (5.21)$$

From this characteristic function (5.21) the distribution of the quadratic forms (5.20) can be derived by means of the inversion theorem. To do so, we need to express the term $\Im\left(e^{-ity} \cdot \phi_y(t)\right)$ in known quantities. Note that $\Im(\cdot)$ is free of any imaginary expressions so that the integration in equation (5.19) can be performed straight forward in the domain of real numbers.

Let us first recap some properties of complex numbers which are needed later: any complex number $z = a + i \cdot b$ can be written as two-dimensional vector in the complex plane (see Figure 5.1), where a is the real part of z, b its imaginary part, and i the imaginary unit. The *modulus* of this complex vector (i.e., its length) is defined by

$$|z| = \sqrt{a^2 + b^2} \qquad (5.22)$$

and the *argument* of z (i.e., its angle to the real axes) is given by

$$\arg(z) = \varphi = \arctan\left(\frac{b}{a}\right) \qquad (5.23)$$

with $-\pi \leq \arg(z) \leq \pi$. Consequently, the imaginary part $\Im(z) = b$ is given by

$$\Im(z) = \sin\big(\arg(z)\big) \cdot |z| \ . \tag{5.24}$$

By using the trigonometrical form of complex numbers, z can be written as $z = |z| \cdot \big(\cos(\varphi) + i \cdot \sin(\varphi)\big)$. The theorem by De Moivre states that the q-th power of a complex number z for any real q is

$$z^q = |z|^q \big(\cos(q \cdot \varphi) + i \cdot \sin(q \cdot \varphi)\big) \ . \tag{5.25}$$

Finally, the product of two complex numbers z_1 and z_2 is

$$z_1 \cdot z_2 = |z_1| \cdot |z_2| \big(\cos(\varphi_1 + \varphi_2) + i \cdot \sin(\varphi_1 + \varphi_2)\big) \ . \tag{5.26}$$

Thus its modulus is the product of its single moduli (i.e., $|z_1| \cdot |z_2|$) and its argument is the sum of the single arguments, i.e., $\varphi_1 + \varphi_2$.

To express $\Im\big(e^{-ity} \cdot \phi_y(t)\big)$ in terms of the argument $\arg\big(e^{-ity} \cdot \phi_y(t)\big)$ and the modulus $|e^{-ity} \cdot \phi_y(t)|$ let us concentrate first on the single factors $\phi_{y_j}(t) = (1 - 2it\lambda_j)^{-\frac{1}{2}}$ and e^{-ity_j} where use has been made of the definition $y_j \equiv \lambda_j \cdot x_j^2$. We get for the argument

$$\arg\big((1 - 2it\lambda_j)^{-\frac{1}{2}}\big) = -\frac{1}{2}\arctan(-2t\lambda_j) = \frac{1}{2}\arctan(2t\lambda_j) \tag{5.27}$$

and

$$\arg(e^{-ity_j}) = -t \cdot y_j \tag{5.28}$$

as well as for the modulus

$$|(1 - 2it\lambda_j)^{-\frac{1}{2}}| = (1 + 4t^2\lambda_j^2)^{-\frac{1}{4}} \tag{5.29}$$

and

$$|e^{-ity_j}| = 1 \ . \tag{5.30}$$

Collecting these results, making use of equation (5.26), and substituting the variable t by $u \equiv 2 \cdot t$ we get

$$\theta(u) \equiv \arg\big(e^{-iuy} \cdot \phi_y(u)\big) = \frac{1}{2}\sum_{j=1}^{n}\arctan(u\lambda_j) - \frac{1}{2}u \cdot y \tag{5.31}$$

and

$$\rho(u) \equiv |e^{-iuy} \cdot \phi_y(u)| = \prod_{j=1}^{n}(1 + u^2\lambda_j^2)^{-\frac{1}{4}} \ . \tag{5.32}$$

Substituting equations (5.31) and (5.32) into $\Im\big(e^{-ity} \cdot \phi_y(t)\big) = \sin\big(\theta(u)\big) \cdot \rho(u)$ and this, in return, into the inversion formula (5.19) the distribution of a weighted sum of independent central χ^2-distributed variables becomes

$$\Pr(Y \leq y) = \frac{1}{2} - \frac{1}{\pi}\int_0^\infty \frac{1}{u} \cdot \sin\big(\theta(u)\big) \cdot \rho(u) \cdot du \ . \tag{5.33}$$

This equation for the distribution of a quadratic form (5.20) is due to Imhof (1961) and will be subsequently called Imhof's formula.

5.4   Extensions and Comments

With increasing computing power the academic interest in the characteristic function to derive an underlying distribution function is emerging, too. For instance, in epidemiology it is used to evaluate the power of focused tests to detect disease clustering around a putative source (Waller and Lawson, 1995, and Waller, Turnbull and Hardin, 1995). In addition, the characteristic function $\phi(t)$ can be extended for random vectors from multivariate distributions (see for instance Stuart and Ord, 1994, p 140ff, and Shephard, 1991a). Also here applications can be found to derive the joint distribution of two ratios in quadratic forms (see Shephard, 1991b and Shively, Ansley and Kohn, 1990). However, as Shephard (1991b) points out, as long as the two matrices underlying both ratios of quadratic forms do not commute evaluation of the joint distribution function can become quite expensive in terms of computing time. Nevertheless, he states that if the characteristic function is reasonably simple and that if the dimension of integration is reasonably small then there are possibilities of large computational economies when compared with simulation experimentation techniques.

While the approaches mentioned above are all based on a real-valued integral of the characteristic function, there are also approaches which directly evaluate the complex-valued characteristic function of a weighted sum of χ^2-distributed variables. However, these complex-valued approaches have not succeeded in the daily practice because they have to take singularities at the weights λ_i into account and are too costly to implement. Notable here the work by Pan (1968) has to be mentioned and the implementation of his formulas in a computer program by Farebrother (1980, 1981, and 1984a). Also the Fourier inversion and the fast Fourier transformation are proposed in the literature to derive the distribution function from general characteristic functions which need not be a weighted sum of χ^2-distributed variables. These methods and their associated problems are discussed in Bohman (1970, 1972, 1975, and 1980), Davis and Borgman (1979), Good (1987) and Diggle and Hall (1993). How far the numeric Fourier inversion is capable of satisfactory approximating the distribution function still needs to be investigated further.

Chapter 6

Numerical Evaluation of Imhof's Formula

This section will discuss the properties of Imhof's formula and gives suggestions how to evaluate it numerically. The most critical entity in Imhof's formula is the spectrum of weights $\{\lambda_i\}$ and their distribution around zero. They determine the distribution function and will consequently be discussed first. Then a short numerical example will outline some of the properties associated with Imhof's formula and this section highlights a problem attributable to the improper integral (5.33) which can not be evaluated up to $u \to \infty$. At the other end also the behavior at $u = 0$ deserves some attention.

6.1 Influential Weights in Imhof's Formula

Because all central χ^2-distributed random variables X_i share the same distribution the only component differentiating the distributions of the weighted random sums Ys are their underlying spectra of weights $\{\lambda_j\}$. Weights which equal zero (i.e., $\lambda_j = 0$) do not have an impact on Imhof's formula and can therefore be ignored. This can be best seen by noting that $\arctan(u \cdot 0) = 0$ is a neutral summand in the summation term $\theta(u)$ and that $(1 + u^2 \cdot 0) = 1$ is the neutral factor in the product term $\rho(u)$. Therefore, assuming m weights with $m \le n$ are non-zero, the remaining zero $n - m$ weights can be dropped from the spectrum. Consequently, calculation can be performed solely with the re-sequenced set of non-zero weights $\{\lambda_j, j = 1, \ldots, m\}$. Especially, when there is a large number of zero weights, calculation of the distribution function based on reduced spectrum of weights translates into a substantial saving of computing time and resources. Furthermore, following the same line of argumentation, larger absolute weights $|\lambda_j|$ in the spectrum of weights have a stronger impact on the distribution of the weighted sum Y than smaller absolute weights.

6.2 A Numerical Example

Figure 6.1 provides some insight into the behavior of the integrand $f(u) = \frac{1}{u} \cdot \sin(\theta(u)) \cdot \rho(u)$ for a set of 3 weights $\boldsymbol{\lambda} = (1, 2, 3)^T$. Since χ^2-distributed variables are positive and because all weights in this example are greater than zero, the weighted sum Y has to be greater equal than zero. It can be seen that the integrand is a well-behaved continuous function and that it converges to zero with increasing u. The integration to derive the distribution for a given observation $Y \le y$ is performed along the u-axis displayed in the foreground. Notice that in the region around $Y \gtrsim 0$ the function $\frac{1}{u} \cdot \sin(\theta(u)) \cdot \rho(u)$ fades out

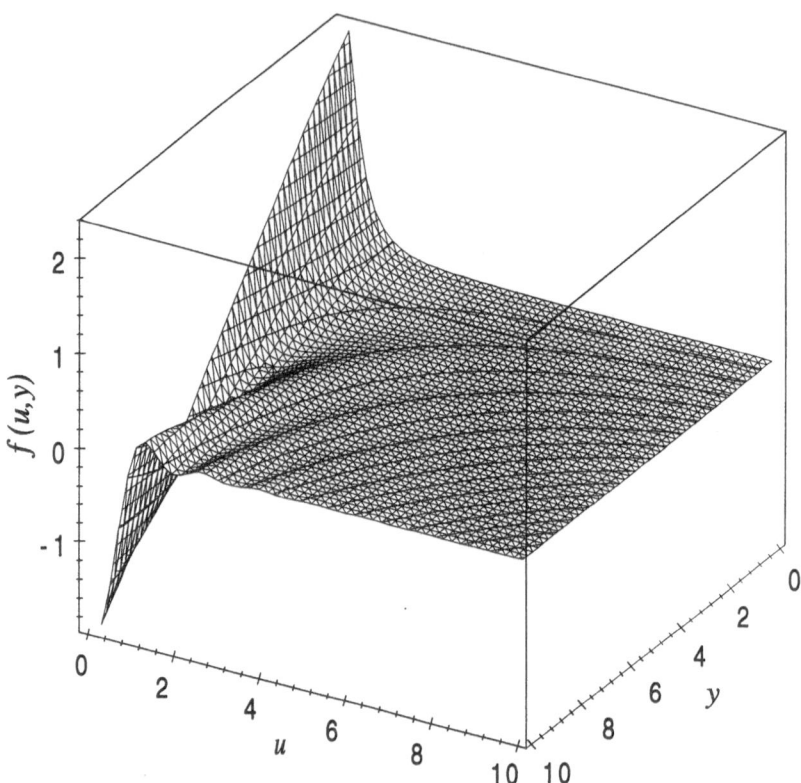

Fig. 6.1. Behavior of the integrand $\frac{1}{u} \cdot \sin\big(\theta(u)\big) \cdot \rho(u)$ for the weights $\lambda = (1,2,3)^T$ in Imhof's formula

rapidly along the u-axis. In contrast, at around $Y \approx 10$ the integrand oscillates with dampening amplitude around the zero. Furthermore, for $u = 0$ a linear trend in the integrand can be observed over the y-axes. At $y = 0$ the integrand is positive over the complete u-axis and equals $\pi/2$ so that $\Pr(Y \leq 0) = 0$.

6.3   The Behavior of the Integrand at the Origin

Let us first concentrate on the behavior of the integrand $\frac{1}{u} \cdot \sin\big(\theta(u)\big) \cdot \rho(u)$ at $u = 0$. Imhof (1961) gives the limit as u approaches zero

$$\lim_{u \to 0} \frac{1}{u} \cdot \sin\big(\theta(u)\big) \cdot \rho(u) = \frac{1}{2} \sum_{j=1}^{m} \lambda_j - \frac{1}{2} \cdot y \tag{6.1}$$

which is a linear function in the observed value y as can be seen in Figure 6.1. This value at $u = 0$ can be used as starting value for numerical integration algorithms with variable step size. In contrast, the next section will deal with the behavior of the integrand as u approaches infinity.

6.4 Derivation of an Upper Truncation Point

The previous example raised the question at which value of u for a given y the integration process can be stopped so that the truncation error of the improper integral (5.33) is negligible. Imhof (1961) also assesses this truncation error t_U which measures the inaccuracy if we stop evaluating the integral at a given value T_U with $0 \le u \le T_U < \infty$. It is defined by

$$t_U \equiv \frac{1}{\pi} \int_{T_U}^{\infty} \frac{1}{u} \cdot \sin\big(\theta(u)\big) \cdot \rho(u) \cdot du \tag{6.2}$$

This truncation error t_U can be bounded from above as follows:

$$|t_U| = \frac{1}{\pi} \left| \int_{T_U}^{\infty} \frac{1}{u} \cdot \sin\big(\theta(u)\big) \cdot \rho(u) \cdot du \right| \tag{6.3}$$

$$\le \frac{1}{\pi} \int_{T_U}^{\infty} \left| \frac{1}{u} \cdot \sin\big(\theta(u)\big) \cdot \rho(u) \right| \cdot du$$

$$\le \frac{1}{\pi} \int_{T_U}^{\infty} \left| \frac{1}{u} \cdot \rho(u) \right| \cdot du \ .$$

Thus we need to concentrate only on $\rho(u)$ and remember that $u \ge 0$. Then the inequality

$$(1 + u^2 \lambda_j^2)^{-\frac{1}{4}} < (u^2 \lambda_j^2)^{-\frac{1}{4}} = u^{-\frac{1}{2}} |\lambda_j|^{-\frac{1}{2}} \tag{6.4}$$

holds and therefore we can write

$$\prod_{j=1}^{m} (1 + u^2 \lambda_j^2)^{-\frac{1}{4}} < \prod_{j=1}^{m} u^{-\frac{1}{2}} |\lambda_j|^{-\frac{1}{2}} \ . \tag{6.5}$$

Consequently the upper bound of the truncation error t_U can be written as

$$|t_U| < \frac{1}{\pi} \int_{T_U}^{\infty} \frac{1}{u} \cdot \prod_{j=1}^{m} u^{-\frac{1}{2}} |\lambda_j|^{-\frac{1}{2}} \cdot du \tag{6.6}$$

$$< \frac{1}{\pi} \cdot \prod_{j=1}^{m} |\lambda_j|^{-\frac{1}{2}} \int_{T_U}^{\infty} u^{-\left(\frac{m}{2}+1\right)} \cdot du$$

$$< \frac{1}{\pi} \cdot \prod_{j=1}^{m} |\lambda_j|^{-\frac{1}{2}} \cdot \frac{2}{m} \cdot T_U^{-\frac{m}{2}} \ .$$

Solving for T_U at a pre-defined truncation error t_U finally gives the truncation point

$$T_U = \left(t_U \cdot \pi \cdot \frac{m}{2} \cdot \prod_{j=1}^{m} |\lambda_j|^{\frac{1}{2}} \right)^{-\frac{2}{m}} \tag{6.7}$$

at which the numerical integration can stop with a desired level of accuracy. Since the truncation point T_U is not only undefined for weights equal to zero but also weights close to zero cause numerical problems, it is suggested that Farebrother's (1980) recommendation be followed and that all weights in a small interval around zero be dropped and m be adjusted accordingly.

An alternative method to deal with the truncation problem directly would be to perform a change in the integration variable u for a general function defined over $u \in [0, \infty)$. If the integrand $f(u)$ decreases exponentially the following substitution can be performed (see Press et al., 1989, p 137). Let

$$v = \exp(-u) \quad \text{or} \quad u = -\ln(v) \tag{6.8}$$

then

$$\int_{u=0}^{u=\infty} f(u) \cdot du = \int_{v=\exp(-\infty)}^{v=\exp(0)} \frac{1}{v} \cdot f(-\ln(v)) \cdot dv \tag{6.9}$$

with the new lower bound of the transformed integral being $\exp(-\infty) = 0$ and its upper bound being $\exp(-0) = 1$. The properties of this change in the integration variable u for Imhof's formula still needs to be investigated more closely. For instance, the lower bound $\lim_{v \to 0} \frac{1}{v} f(-\ln(v))$ of the integrand needs to be determined if it exists.

6.5   Numerical Algorithms to Approximate Imhof's Formula

While the previous section discussed the truncation error another error arrises by numerically approximating an integral (5.33). The process of numerical calculation of an integral is also called in the literature quadrature because an integral can be approximated by a series of rectangular panels. The general goal is to come as close as possible the real integral value of a continuous function $f(u)$ by using a minimum number of numerical evaluations of the integrand.

The key idea, dating back to Riemann's (1826–66) sum, is to split up the integral into a discrete summation over k weighted function values $f(u_i)$. I.e., let $f(u)$ be the continuous function under investigation in the interval $[a, b]$ then we are looking for an approximation such as

$$\int_a^b f(u) \cdot du \approx \sum_{i=1}^{k} f(u_i) \cdot w_i \tag{6.10}$$

where w_i are the weights such that $\sum_{i=1}^{k} w_i = b - a$. The algorithms to numerically approximate the integral $\int_a^b f(u) \cdot du$ are distinguished by the choice of the weights w_i on which the values u_i on the abscissa also depend.

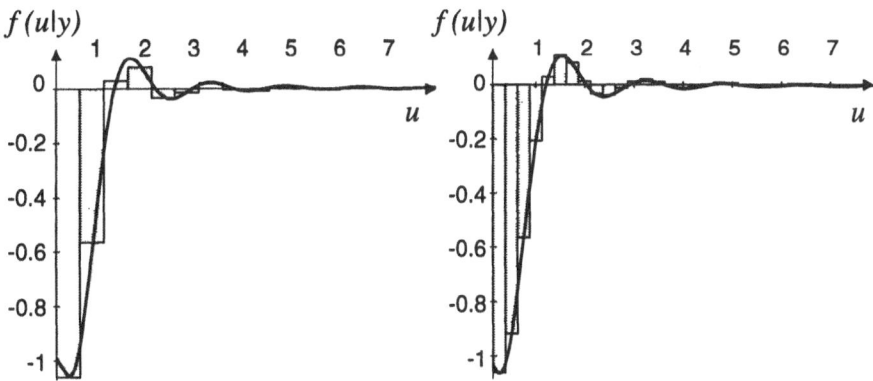

Fig. 6.2. Evaluation of a Riemann sum $\frac{1}{u} \cdot \sin(\theta(u)) \cdot \rho(u)$ with weights $\lambda = (1, 2, 3)^T$ and increasing function evaluations

6.5.1   Equally Spaced Step-Size

The simplest rational to evaluate (6.10) is to select constant weights w_i such that $w = (b - a)/k$ and determine the arguments u_i on the abscissa by $u_i = a + i \cdot w - w/2$. By increasing the number of intervals k or, respectively, reducing step-size w the precision of the approximation in (6.10) must increase. An elegant choice to increase the number of intervals is to proceed successively in potencies of 2. This way the function values at $f(u_i^j)$ from the previous step j can be reused by interleaving these with the new intermediate values.

Figure 6.2 displays the underlying idea of this successive step-halving for the example given in Section 6.2 with the truncation point set to $T_U = 8$. With decreasing step-size the precision of the approximation increases. The left-hand figure displays 16 intervals in the range $[0, 8]$. Note that for $y = 8$ the value $\lim_{u \to 0} \frac{1}{u} \cdot \sin(\theta(u)) \cdot \rho(u) = 1$ which is in-line with equation (6.1). In contrast, to allow the interleaving half-step algorithm to work properly, the right-hand figure displays only 31 intervals in the range $[0.125, 7.875]$. It is evident that an adjustment in the step-size and in the weights is needed at the endpoints of the integral. For a detailed discussion see for instance Press et al. (1989, p 119).

Throughout all calculations I have used the adjustment that is implemented in GAUSS (Aptech Inc., 1996). Nevertheless, to save computer memory I needed to modify the implementation of GAUSS's algorithm slightly. GAUSS is a 32-bit programming language which is optimized for large arrays. Therefore, instead of successively adding up the weighted function values $f(u_i)$ it first calculates the full array and stores it in the core memory for subsequent summation. With decreasing step-size the number of function values $f(u_i)$ increases so that the limit of the physical core memory can be reached swiftly, especially when the truncation point T_U is determined to be excessively conservative. I have selected a compromise between speed and memory usage: along the abscissa only an intermediate sub-array with a limited number of $f(u_i)$ evaluations is calculated in the memory and subsequently its elements are summed up into a simple scalar.

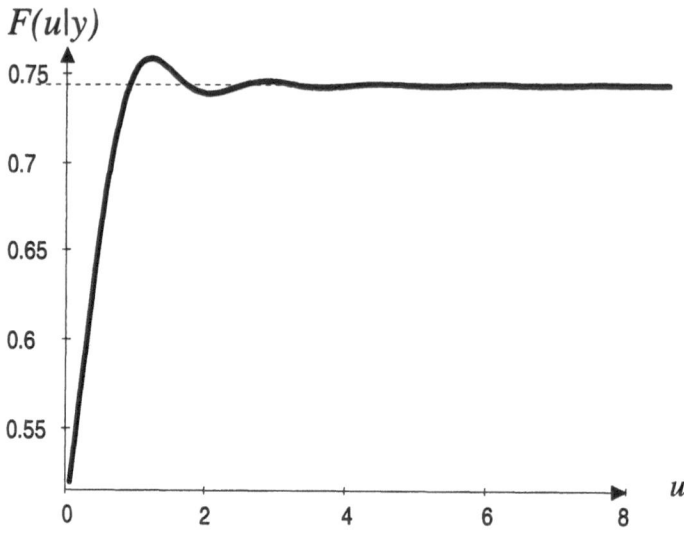

Fig. 6.3. Graphical solution of Imhof's formula expressed as differential equation for $\Pr(Y \le 8)$ and $\boldsymbol{\lambda} = (1, 2, 3)^T$. The abscissa displays u with the truncation point set to $T_U = 8$

Then the next successive sub-array of intermediate function values is processed and its elements are added to the first sum. This process is repeated until the truncation point T_U is reached. In GAUSS the accuracy of the step-halving is determined by a user-defined integration error ϵ which is

$$\left| \frac{\text{new integral} - \text{old integral}}{\text{old integral}} \right| < \epsilon . \tag{6.11}$$

The numerical quadrature terminates after the new integral satisfies the inequality given in equation (6.11).

6.5.2   Variable Spaced Step-Size

Another approach to numerically integrate the characteristic function is to formulate the problem as a differential equation. To solve differential equations rather sophisticated algorithms like the *Bulirsch-Stoer* method are available (see for instance Press et al., 1989, Chap. 15) which approximate an integral with a minimum of function evaluations. To do so, these algorithms handle the step-size adaptively in dependence of the curvature of the integrand. As long as the integrand has a constant slope over a long range along the abscissa it can be very well approximated by a linear function and requires only one evaluation of $f(u)$. Whereas, if the curvature of the integrand is large the step-size has to be decreased to capture its variability. For instance, in Figure 6.2 the slope of the integrand $\frac{1}{u} \cdot \sin(\theta(u)) \cdot \rho(u)$ is virtually constant over the range $[0.4, 1.2]$ and can therefore very well be approximated by only one panel. In contrast, to approximate the integral over the range $[1.2, 1.6]$ several panels are needed.

The probability to observe a weighted sum Y of χ^2-distributed variables which is smaller than 8 for the example given in Section 6.2 is $\Pr(Y \leq 8) = 0.744$. Expressing Imhof's formula in equation (5.33) as first order differential equation with $u \geq 0$ we get

$$\left. \begin{array}{c} \frac{dF(u)}{du} = -\frac{1}{\pi} \cdot \frac{1}{u} \cdot \sin(\theta(u)) \cdot \rho(u) \\ F(0) = \frac{1}{2} \end{array} \right\} \tag{6.12}$$

where $F(0) = \frac{1}{2}$ is the initial value. A numerical solution for this differential equation is displayed in Figure 6.3. It shows the sequential accumulation of the integrand given in Figure 6.2 up to $T_U = 8$. Notice that for $u > 6$ the solution of the differential system stabilizes at $\Pr(Y \leq 8) = 0.744$. This demonstrates that, for the formulation of Imhof's formula as a differential equation, the truncation point does not need to be known. It is worth comparing the performance of variable spaced step-size algorithms against fixed step-size algorithms for Imhof's formula in future investigations.

Chapter 7

The Exact Distribution of Moran's I

This chapter will make extensive use of Imhof's formula to calculate the exact distribution of Moran's I under either the assumption of spatial independence or conditionally on the forces of a spatial process. To do so several steps are required: (i) the regression residuals have to be transformed into independent identically normal distributed variables, (ii) Moran's I, which is a ratio of quadratic forms, has to be converted into the single quadratic form, and (iii) finally the weights λ_i have to be determined for calculation of the weighted sum of χ^2-distributed variables. The pivotal entity in this discussion is the spectrum of eigenvalues from the coded spatial link matrix adjusted by the projection matrix and the impact of a hypothetical spatial process. Because under the assumption of spatial independence the spectrum of eigenvalues bears some invariance properties with respect to the given significance points (i) the calculation of the eigenvalues can be simplified and, (ii) an upper and a lower bound of Moran's I exact distribution can be given for the case that we neglect the impact of design matrix on the spectrum of the eigenvalues.

The knowledge of the exact distribution of Moran's I both under the zero hypothesis and under alternative hypotheses provides a benchmark to assess its proposed approximate normal distribution and other asymptotic distribution models. Moreover, it reduces the need for simulation studies in many circumstances. The methods outlined here allow me to address many questions raised in the literature and endows applicable solutions to new problems not yet considered. Besides the work by Imhof, for much of the present discussion the seminal paper by Durbin and Watson (1950) has to be credited. Finally, Koerts and Abrahamse's (1969) knowledgable merger of Imhof's formula with Durbin and Watson results allowed for the first time to give the exact distribution of a general ratio of quadratic forms. A discussion on the exact moments of Moran's I is deferred until Chapter 9.

7.1 Derivation of the Distribution under the Influence of a Spatial Process

Recall that the observed value of Moran's I_0 is defined as a ratio of quadratic forms in the regression residuals $\hat{\varepsilon}$ by

$$I_0 = \frac{\hat{\varepsilon}^T \cdot \frac{1}{2} \cdot (\mathbf{V} + \mathbf{V}^T) \cdot \hat{\varepsilon}}{\hat{\varepsilon}^T \cdot \hat{\varepsilon}} \tag{7.1}$$

where

$$\widehat{\varepsilon} = \mathbf{M} \cdot \mathbf{y} \quad \text{or equivalently} \quad \widehat{\varepsilon} = \mathbf{M} \cdot \varepsilon . \tag{7.2}$$

This statistic is of the same structure as the Durbin and Watson's d-statistic (1950) for serial autocorrelation. The only difference lies in the symmetric structure matrix $\frac{1}{2} \cdot (\mathbf{V} + \mathbf{V}^T)$ which, in the case of spatial autocorrelation, reflects a hypothetical multilateral mutually dependency structure between the spatial objects and has to be specified on theoretical grounds. Consequently, all theoretical results on Durbin and Waston's d-statistic can be easily transferred to Moran's I.

7.1.1   Transformation into Independent Random Variables

Under the presence of a significant spatial process the disturbances follow a normal distribution with

$$\varepsilon \sim \mathcal{N}(0, \sigma^2 \cdot \mathbf{\Omega}) \tag{7.3}$$

so that the regression residuals have a co-variance matrix determined by $\text{Cov}[\widehat{\varepsilon}] = \text{E}[\mathbf{M} \cdot \varepsilon \cdot \varepsilon^T \cdot \mathbf{M}]$ and are normal distributed with

$$\widehat{\varepsilon} \sim \mathcal{N}(0, \sigma^2 \cdot \mathbf{M} \cdot \mathbf{\Omega} \cdot \mathbf{M}) . \tag{7.4}$$

Note that the expectation of the residuals $\widehat{\varepsilon}$ is zero; this is important in so far only as this leads to *central* χ^2-distributed variables. However, the distribution of the regression residuals $\widehat{\varepsilon}$ also depends on the design matrix $\mathbf{X}$ by means of the projection matrix $\mathbf{M}$. Furthermore, let $\boldsymbol{\delta}$ be a random spherically symmetric distributed vector[1] such that $\boldsymbol{\delta} \sim \mathcal{N}(\mathbf{0}, \sigma^2 \cdot \mathbf{I})$. Then the autocorrelated disturbances ε can be generated by a linear transformation

$$\varepsilon = \mathbf{\Omega}^{\frac{1}{2}} \cdot \boldsymbol{\delta} \tag{7.5}$$

of the independent random vector $\boldsymbol{\delta}$. The transformation matrix $\mathbf{\Omega}^{\frac{1}{2}}$ may take many but equivalent forms (see for instance Haining, 1990, p 117): it can be the lower triangular matrix of a Cholesky decomposition of $\mathbf{\Omega}$; it can be derived by an eigensystem decomposition $\mathbf{\Omega} = \mathbf{P} \cdot \mathbf{\Lambda} \cdot \mathbf{P}^T$ where the columns of the matrix $\mathbf{P}$ are the normalized eigenvectors of the co-variance matrix $\mathbf{\Omega}$ and $\mathbf{\Lambda}$ is the diagonal matrix of eigenvalues so that $\mathbf{\Omega}^{\frac{1}{2}} = \mathbf{P} \cdot \mathbf{\Lambda}^{\frac{1}{2}}$; or in case of the plain spatial MA- or AR-processes $\mathbf{\Omega}^{\frac{1}{2}}$ can be either $\sigma \cdot (\mathbf{I} + \rho \cdot \mathbf{V})$ or $\sigma \cdot (\mathbf{I} - \rho \cdot \mathbf{V}^T)^{-1}$, respectively.

---

[1] A random vector is spherically symmetric if its distribution laws are invariant to rotations about the origin (i.e., to orthogonal transformations). $\mathcal{N}(\mathbf{0}, \sigma^2 \cdot \mathbf{I})$ is an example of such a distribution (see King, 1981).

7.1.2   Transformation of Moran's I into a Simple Quadratic Form

Collecting previous transformations (7.2) and (7.5) of the regression residuals, Moran's I can be written in terms of the unknown independent random vector δ, i.e.,

$$I_0 = \frac{\delta^T \cdot \Omega^{T\frac{1}{2}} \cdot \mathbf{M} \cdot \frac{1}{2} \cdot (\mathbf{V} + \mathbf{V}^T) \cdot \mathbf{M} \cdot \Omega^{\frac{1}{2}} \cdot \delta}{\delta^T \cdot \Omega^{T\frac{1}{2}} \cdot \mathbf{M} \cdot \Omega^{\frac{1}{2}} \cdot \delta} . \tag{7.6}$$

Hence, recalling that the projection matrix $\mathbf{M}$ is idempotent, the conditional distribution $F(I_0 \mid \Omega)$ of Moran's I given the observed value I_0 and a hypothetical spatial process Ω can be written as

$$F(I_0 \mid \Omega)$$

$$= \Pr\left(\frac{\delta^T \cdot \Omega^{T\frac{1}{2}} \cdot \mathbf{M} \cdot \frac{1}{2} \cdot (\mathbf{V} + \mathbf{V}^T) \cdot \mathbf{M} \cdot \Omega^{\frac{1}{2}} \cdot \delta}{\delta^T \cdot \Omega^{T\frac{1}{2}} \cdot \mathbf{M} \cdot \Omega^{\frac{1}{2}} \cdot \delta} \leq I_0 \right), \tag{7.7}$$

$$= \Pr\left(\delta^T \Omega^{T\frac{1}{2}} \mathbf{M} \cdot \frac{1}{2}(\mathbf{V} + \mathbf{V}^T) \cdot \mathbf{M}\Omega^{\frac{1}{2}}\delta \leq I_0 \cdot \delta^T \Omega^{T\frac{1}{2}} \mathbf{M}\Omega^{\frac{1}{2}}\delta \right), \tag{7.8}$$

$$= \Pr\left(\delta^T \cdot \Omega^{T\frac{1}{2}} \cdot \mathbf{M} \cdot \left[\frac{1}{2}(\mathbf{V} + \mathbf{V}^T) - I_0 \cdot \mathbf{I} \right] \cdot \mathbf{M} \cdot \Omega^{\frac{1}{2}} \cdot \delta \leq 0 \right). \tag{7.9}$$

This algebraic transformation to bring a ratio of quadratic forms into a simple quadratic form was proposed by Koerts and Abrahamse (1968). However, the new random variable on the left hand-side depends on the observed level I_0 and its probability is always evaluated up to 0.

7.1.3   Determination of the Weights

Because the random vector δ belongs to the class of the spherically symmetric distributions (see King, 1980) it can be transformed by an orthogonal transformation matrix $\mathbf{H}$ such that

$$\eta = \mathbf{H} \cdot \delta \tag{7.10}$$

is again independently normally distributed with $\eta \sim \mathcal{N}(0, \sigma^2 \cdot \mathbf{I})$. If we define the orthogonal transformation matrix $\mathbf{H}$ by the normalized eigenvectors[2] of the term $\Omega^{T\frac{1}{2}} \cdot \mathbf{M} \cdot [\frac{1}{2}(\mathbf{V} + \mathbf{V}^T) - I_0 \cdot \mathbf{I}] \cdot \mathbf{M} \cdot \Omega^{\frac{1}{2}}$ from equation (7.9) with the diagonal eigenvalue matrix Λ then the conditional distribution of Moran's I can be written as

$$F(I_0 \mid \Omega) = \Pr(\eta^T \cdot \Lambda \cdot \eta \leq 0), \tag{7.11}$$

$$= \Pr\left(\sum_{i=1}^{n} \lambda_i \cdot \eta_i^2 \leq 0 \right). \tag{7.12}$$

---

[2] Note that if the spatial link matrix is not symmetric or transformed to symmetry the matrix $\mathbf{H}$ would not be orthogonal. Therefore, a transformation of the disturbances by a non-orthogonal matrix $\mathbf{H}$ ends up in non-independent random variables.

The term $\sum_{i=1}^{n} \lambda_i \cdot \eta_i^2$ is a weighted sum of the χ^2-distributed variables η_i^2. Consequently, Imhof's formula can be applied on equation (7.12) to calculate the distribution of Moran's I. Recall that zero eigenvalues λ_i are permitted and that the distribution function $F(I_0 \mid \boldsymbol{\Omega})$ is invariant under their presence (see Chap. 6.1).

7.1.4   Discussion

Several comments have to be made about equation (7.12):

(i) The spectrum of eigenvalues in $\boldsymbol{\Lambda}$ not only depends on the spatial link matrix $\mathbf{V}$ but also on the observed level I_0, the projection matrix $\mathbf{M}$ and the hypothetical spatial process characterized by the co-variance matrix $\boldsymbol{\Omega}$. It therefore has to be re-evaluated if any of these four components changes (see Section 7.2 for a simplification). In particular, if we want to specify the complete distribution function $F(I_0 \mid \boldsymbol{\Omega})$ of Moran's I under a given model we need to evaluate the spectrum of eigenvalues for a large number of significance points I_0 within the feasible range of Moran's I.

(ii) By definition the co-variance matrix $\boldsymbol{\Omega}$ is strong positive definite (see Section 4.2 for the feasible range of autocorrelation level ρ) whereas the projection matrix $\mathbf{M}$ has only a rank of $n-k$. The rank of the spatial link matrix $\mathbf{V}$ depends on its definition. For regular spatial lattices (i.e., triangles, rectangles, or hexagons) and usually for large planar spatial structures $\mathbf{V}$ does not have full rank. Consequently, due to the rank defect of the projection matrix at least k eigenvalues of $\boldsymbol{\Lambda}$ are zero and not relevant for the calculation of the distribution of Moran's I.

(iii) Only the distribution of the eigenvalues in $\boldsymbol{\Lambda}$ determines the shape of the distribution function of Moran's I.

(iv) In terms of the computing time it is in general more expensive to calculate the spectrum of eigenvalues $\boldsymbol{\Lambda}$ than to perform subsequently the numerical integration to derive $F(I_0 \mid \boldsymbol{\Omega})$ at one significance point I_0. The calculation of eigenvalues requires $O(n^3)$ operations (Shively, Ansley, and Kohn, 1990). In contrast, for a given smoothly distributed spectrum of eigenvalues the computation of the exact probabilities becomes accelerated when moving from small n (say < 10) to an intermediate n (say ≈ 50) and then slightly increases with order $O(n)$.

(v) Due to Koerts and Abrahamse's transformation of Moran's I to a simple weighted sum, Imhof's formula is always evaluated at 0 so that the term $-\frac{1}{2}u \cdot y$ in $\theta(u)$ is irrelevant (see equation 5.31).

Besides the procedure outlined above to calculate the distribution of Moran's I *eigenvalue-free* methods have also been suggested in the literature. These methods are based on the direct evaluation of the characteristic function

(5.21) in terms of the determinant[3]

$$\phi_y(t) = \left| \mathbf{I} - 2it \cdot \left(\mathbf{\Omega}^{T\frac{1}{2}} \cdot \mathbf{M} \cdot \left[\frac{1}{2}(\mathbf{V} + \mathbf{V}^T) - I_0 \cdot \mathbf{I} \right] \cdot \mathbf{M} \cdot \mathbf{\Omega}^{\frac{1}{2}} \right) \right|^{-\frac{1}{2}} \quad (7.13)$$

which leads, for special band diagonal forms of $\mathbf{V}$ and $\mathbf{\Omega}$, to a substantial saving in computing time because the calculation of the eigenvalues can be skipped. Usually, in the time-series situation the link matrix $\mathbf{V}$ is band diagonal and under the zero hypothesis the covariance matrix is diagonal (i.e., $\mathbf{\Omega} \equiv \sigma^2 \cdot \mathbf{I}$). Palm and Sneek (1984) and Farebrother (1985 and 1990) use this approach after transforming $\mathbf{M} \cdot [\frac{1}{2}(\mathbf{V} + \mathbf{V}^T) - I_0 \cdot \mathbf{I}] \cdot \mathbf{M}$ to a tridiagonal form for the Durbin and Watson d-statistic under the zero hypothesis. Under the zero hypothesis for the time-series situation Shively, Ansley, and Kohn (1990), Ansely, Kohn, and Shively (1992), and Kohn, Shively, and Ansely (1993) push this eigenvalue-free method in conjunction with a modified Kalman filter and a Cholesky decomposition further and gain a substantial saving in computing time. However, as Farebrother (1994) points out their approach is only applicable for central χ^2-distributed variables and matrices in band form, and that in special circumstances the Cholesky's procedure is numerically unstable. In the spatial situation, however, I feel that the eigenvalue-free approach is not feasible because usually spatial structure matrices are by far more complex than link matrices in the time-series case.

7.2   Simplifications under the Assumption of Spatial Independence

The matrices $\mathbf{\Omega}$, $\mathbf{M}$, $\mathbf{V}$ are invariant for a given problem under investigation. However, we can see from the term $\mathbf{\Omega}^{T\frac{1}{2}} \cdot \mathbf{M} \cdot [\frac{1}{2}(\mathbf{V} + \mathbf{V}^T) - I_0 \cdot \mathbf{I}] \cdot \mathbf{M} \cdot \mathbf{\Omega}^{\frac{1}{2}}$ that the spectrum of eigenvalues needs to be evaluated anew for each observed significance point I_0. If we want to calculate the full distribution function and not just the probability of one I_0 this can become a quite time consuming task for larger spatial settings. If however the co-variance matrix $\mathbf{\Omega} \equiv \sigma^2 \cdot \mathbf{I}$ (i.e., we make use of the assumption of spatial independence) then previous term reduces to $\mathbf{M} \cdot [\frac{1}{2}(\mathbf{V} + \mathbf{V}^T) - I_0 \cdot \mathbf{I}] \cdot \mathbf{M}$ for which the spectrum of eigenvalues can be expressed as a linear function of the observed significance point I_0.

However, the distribution of the regression residuals still depends on the projection matrix $\mathbf{M}$ and they follow a distribution $\widehat{\varepsilon} \sim \mathcal{N}(0, \sigma^2 \cdot \mathbf{M})$ where the co-variance is derived by $\mathrm{Cov}[\widehat{\varepsilon}] = \mathrm{E}[\mathbf{M} \cdot \varepsilon \cdot \varepsilon^T \cdot \mathbf{M}] = \mathbf{M} \cdot \mathrm{E}[\varepsilon \cdot \varepsilon^T] \cdot \mathbf{M} = \sigma^2 \cdot \mathbf{M}$ due to the idempotency of the projection matrix $\mathbf{M}$. Therefore, even under the assumption that the invisible disturbances ε are spatially independent the regression residuals $\widehat{\varepsilon}$ are not independent. This problem has caught quite some attention in the econometric literature because the properties of the Durbin and Watson statistic depend crucially on the composition of the design matrix $\mathbf{X}$

---

[3] Note that for any real symmetric matrix $\mathbf{A}$ with eigenvalues $\{\lambda_1, \ldots, \lambda_n\}$ the equivalence $|\mathbf{A}| = \prod_{i=1}^{n} \lambda_i$ holds.

(see King, 1987, for a detailed review). As noted in Chapter 4.3.2 a test, defined as a ratio of quadratic forms in regression residuals and a structure linking these residuals, is a uniformly most powerful test if the design matrix $\mathbf{X}$ lies totally in the vector space spanned by the eigenvectors of the linking structure.

The aim of the following two subsections is to develop a relationship between the spectrum of eigenvalues and the observed I_0 so that the spectrum has to be evaluated only once. The final subsection discusses the relation between the eigenvalues of $\frac{1}{2}(\mathbf{V} + \mathbf{V}^T)$ and $\mathbf{M} \cdot [\frac{1}{2}(\mathbf{V} + \mathbf{V}^T) - I_0 \cdot \mathbf{I}] \cdot \mathbf{M}$ which gives rise to the bounding distributions of Moran's I and is of theoretical interest.

7.2.1   Independence of the Eigenvalue Spectrum from the Denominator

This subsection rests on the initial derivation by Durbin and Watson (1950, best see the corrected reprint in Johnson and Kotz, 1992). Let me argue temporarily in terms of any link matrix $\mathbf{U}$ which either can be real symmetric or real asymmetric. The distinction between symmetric and asymmetric matrices allows me to show that for the derivation of the statistical properties of Moran's I it is mandatory for $\mathbf{U}$ to be symmetric. Alternatively, if $\mathbf{U}$ is not symmetric, it must be forced to symmetry by the transformation $f(\mathbf{U}) \rightarrow \frac{1}{2}(\mathbf{U} + \mathbf{U}^T)$. Furthermore, a corollary of the subsequent discussion is that k eigenvalues must be zero but any eigenvalues beyond these necessarily zero eigenvalues are valid values.

Recall that an observed Moran's I_0 is calculated in terms of the regression residual $\widehat{\varepsilon} \equiv \mathbf{M} \cdot \varepsilon$. Let

$$I_0 = \frac{\varepsilon^T \cdot \mathbf{M} \cdot \mathbf{U} \cdot \mathbf{M} \cdot \varepsilon}{\varepsilon^T \cdot \mathbf{M} \cdot \varepsilon} \tag{7.14}$$

with $\varepsilon_i \sim N(0, \sigma^2)$ being under the zero hypothesis the unobservable, independent disturbances and $\mathbf{M}$ the idempotent projection matrix. Then there is an orthogonal matrix $\mathbf{L}$ (i.e., $\mathbf{I} = \mathbf{L} \cdot \mathbf{L}^T$) such that the kernel of the denominator can be written as

$$\mathbf{L}^T \cdot \mathbf{M} \cdot \mathbf{L} = \begin{pmatrix} \mathbf{I}_{n-k} & \mathbf{0} \\ \mathbf{0} & \mathbf{0}_{k \times k} \end{pmatrix} \tag{7.15}$$

where $\mathbf{I}_{n-k}$ is the identity matrix of order $n - k$ and $\mathbf{0}_{k \times k}$ is a $k \times k$ square matrix of zeros. Equation (7.15) reflects a general property of projection matrices.

Symmetric $\mathbf{U}$: Applying $\mathbf{L}$ simultaneously onto the numerator, its kernel can be re-written as

$$\mathbf{L}^T \cdot \mathbf{M} \cdot \mathbf{U} \cdot \mathbf{M} \cdot \mathbf{L}$$
$$= \mathbf{L}^T \cdot \mathbf{M} \cdot \mathbf{L} \cdot \mathbf{L}^T \cdot \mathbf{U} \cdot \mathbf{L} \cdot \mathbf{L}^T \cdot \mathbf{M} \cdot \mathbf{L} \tag{7.16}$$
$$= \begin{pmatrix} \mathbf{I}_{n-k} & \mathbf{0} \\ \mathbf{0} & \mathbf{0}_{k \times k} \end{pmatrix} \begin{pmatrix} \mathbf{B}_1 & \mathbf{B}_2 \\ \mathbf{B}_3 & \mathbf{B}_4 \end{pmatrix} \begin{pmatrix} \mathbf{I}_{n-k} & \mathbf{0} \\ \mathbf{0} & \mathbf{0}_{k \times k} \end{pmatrix} \tag{7.17}$$
$$= \begin{pmatrix} \mathbf{B}_1 & \mathbf{0} \\ \mathbf{0} & \mathbf{0}_{k \times k} \end{pmatrix} \tag{7.18}$$

where $\left(\begin{smallmatrix} \mathbf{B}_1 & \mathbf{B}_2 \\ \mathbf{B}_3 & \mathbf{B}_4 \end{smallmatrix}\right)$ is the appropriate partitioning of the symmetric matrix $\mathbf{L}^T \cdot \mathbf{U} \cdot \mathbf{L}$. Let $\mathbf{N}_1$ be a matrix diagonalizing $\mathbf{B}_1$, i.e.,

$$\mathbf{N}_1^T \cdot \mathbf{B}_1 \cdot \mathbf{N}_1 = \begin{pmatrix} \lambda_1 & & & \mathbf{0} \\ & \lambda_2 & & \\ & & \ddots & \\ \mathbf{0} & & & \lambda_{n-k} \end{pmatrix}, \tag{7.19}$$

with $\mathbf{N}_1$ being orthogonal. Let us expand $\mathbf{N}_1$ by a $k \times k$ identity matrix $\mathbf{I}_{k \times k}$ and define $\mathbf{N} = \left(\begin{smallmatrix} \mathbf{N}_1 & \mathbf{0} \\ \mathbf{0} & \mathbf{I}_{k \times k} \end{smallmatrix}\right)$ which is again orthogonal. Because the product of two orthogonal matrices is again orthogonal, the product $\mathbf{H} \equiv \mathbf{L} \cdot \mathbf{N}$ is orthogonal.[4] Summing up, we can re-write the kernel of the numerator as

$$\mathbf{\Lambda} = \mathbf{H}^T \cdot \mathbf{M} \cdot \mathbf{U} \cdot \mathbf{M} \cdot \mathbf{H} \tag{7.20}$$

$$= \mathbf{H}^T \cdot \mathbf{M} \cdot \mathbf{L} \cdot \mathbf{L}^T \cdot \mathbf{U} \cdot \mathbf{L} \cdot \mathbf{L}^T \cdot \mathbf{M} \cdot \mathbf{H} \tag{7.21}$$

$$= \mathbf{N}^T \cdot \begin{pmatrix} \mathbf{B}_1 & \mathbf{0} \\ \mathbf{0} & \mathbf{0}_{k \times k} \end{pmatrix} \cdot \mathbf{N} \tag{7.22}$$

$$= \begin{pmatrix} \lambda_1 & & & & \\ & \lambda_2 & & \mathbf{0} & \\ & & \ddots & & \\ & \mathbf{0} & & \lambda_{n-k} & \\ & & & & \mathbf{0}_{k \times k} \end{pmatrix}. \tag{7.23}$$

Consequently, there exists a transformation matrix $\mathbf{H}$ which diagonalizes the numerator for symmetric spatial link matrices $\mathbf{U}$. The question which arises is, whether the matrix $\mathbf{H}$ also diagonalizes the denominator? We can express the denominator for the orthogonal transformation matrix $\mathbf{H}$ as

$$\mathbf{H}^T \cdot \mathbf{M} \cdot \mathbf{H} = \mathbf{N}^T \cdot \mathbf{L}^T \cdot \mathbf{M} \cdot \mathbf{L} \cdot \mathbf{N} \tag{7.24}$$

$$= \mathbf{N}^T \cdot \begin{pmatrix} \mathbf{I}_{n-k} & \mathbf{0} \\ \mathbf{0} & \mathbf{0}_{k \times k} \end{pmatrix} \cdot \mathbf{N} \tag{7.25}$$

$$= \begin{pmatrix} \mathbf{I}_{n-k} & \mathbf{0} \\ \mathbf{0} & \mathbf{0}_{k \times k} \end{pmatrix} \tag{7.26}$$

because $\mathbf{N}$ is orthogonal. Therefore, there exists an orthogonal transformation matrix $\mathbf{H}$ which jointly transforms the random components in the numerator and denominator (i.e., $\varepsilon = \mathbf{H} \cdot \eta$). Applying this transformation, Moran's I can be written in terms of the random variables η_i as

$$I_0 = \frac{\sum_{i=1}^{n-k} \lambda_i \cdot \eta_i^2}{\sum_{i=1}^{n-k} \eta_i^2}. \tag{7.27}$$

where the numerator and denominator are simultaneously diagonalized.

---

[4] This orthogonality can be seen by noting that $\mathbf{H} \cdot \mathbf{H}^T = \mathbf{L} \cdot \mathbf{N} \cdot \mathbf{N}^T \cdot \mathbf{L}^T = \mathbf{L} \cdot \mathbf{I} \cdot \mathbf{L}^T = \mathbf{I}_{n \times n}$.

Asymmetric $\tilde{\mathbf{U}}$: For an asymmetric spatial link matrix $\tilde{\mathbf{U}}$ also the matrix $\tilde{\mathbf{B}}_1 = \mathbf{L}^T \cdot \tilde{\mathbf{U}} \cdot \mathbf{L}$ in (7.19) is not symmetric. Nevertheless, $\begin{pmatrix} \tilde{\mathbf{B}}_1 & 0 \\ 0 & 0_{k \times k} \end{pmatrix}$ can be diagonalized by a transformation matrix $\tilde{\mathbf{N}}$

$$\tilde{\mathbf{\Lambda}} = \tilde{\mathbf{N}}^T \cdot \begin{pmatrix} \tilde{\mathbf{B}}_1 & 0 \\ 0 & 0_{k \times k} \end{pmatrix} \cdot \tilde{\mathbf{N}} , \tag{7.28}$$

as long as the eigensystem remains in the real domain. This is for instance a special property of the row-sum standardized spatial link matrix $\mathbf{W}$ but does not need to hold in general for asymmetric spatial link matrices $\tilde{\mathbf{U}}$. However, the eigenvectors of $\tilde{\mathbf{N}}$ are no longer orthogonal (see Healy, 1992, p 63). Therefore, the denominator can not be reduced to the identity matrix $\mathbf{I}_{n-k}$, i.e.,

$$\tilde{\mathbf{N}}^T \cdot \mathbf{L}^T \cdot \mathbf{M} \cdot \mathbf{L} \cdot \tilde{\mathbf{N}} \neq \begin{pmatrix} \mathbf{I}_{n-k} & 0 \\ 0 & 0_{k \times k} \end{pmatrix} , \tag{7.29}$$

by the same transformation which has been applied onto the numerator.

The existence of a transformation matrix $\mathbf{H}$ which simultaneously diagonalizes the numerator and denominator of Moran's I is a fundamental requirement for the validity of the Durbin and Watson lemma (see Tiefelsdorf and Boots, 1996) which lays the foundation for Moran's I's distribution under the zero hypothesis. Consequently, it only holds for symmetric spatial link matrices.[5]

Erroneously, when dealing with the row-sum standardized spatial link matrix $\mathbf{W}$, Tiefelsdorf and Boots (1995) have ignored in their calculation the transformation $\frac{1}{2}(\mathbf{W} + \mathbf{W}^T)$. This has been corrected in Tiefelsdorf and Boots (1996). However, as a consequence also their conclusion (p 996) that the constant regression vector $\mathbf{1}$ coincides with an eigenvector of the spatial link matrix no longer holds because $\frac{1}{2}(\mathbf{W} + \mathbf{W}^T)$ does not possess a constant eigenvector.

7.2.2 Shifts in the Eigenvalue Spectrum in Dependence of Moran's I's Observed Value

Equation (7.27) states part (a) of the *Durbin and Watson Lemma* and is valid under the assumption of stochastic independence between the regression disturbances. Following the same line of argumentation as in Section 7.1.2 the distribution function of Moran's I can be written as

$$F(I_0 \mid \sigma^2 \cdot \mathbf{I}) = \Pr\left(\frac{\sum_{i=1}^{n-k} \lambda_i \cdot \eta_i^2}{\sum_{i=1}^{n-k} \eta_i^2} \leq I_0 \right) \tag{7.30}$$

$$= \Pr\left(\sum_{i=1}^{n-k} (\lambda_i - I_0) \cdot \eta_i^2 \leq 0 \right) \tag{7.31}$$

---

[5] Ignore the misprint in Johnson and Kotz (1992) on p 241 and read "real symmetric matrix" as stated in the original paper by Durbin and Watson (1950, p 412).

where $\{\lambda_1, \ldots, \lambda_{n-k}, \underbrace{0, \ldots, 0}_{k \text{ zeros}}\}$ is the re-sequenced spectrum of eigenvalues of $\mathbf{M} \cdot$

$\frac{1}{2}(\mathbf{V} + \mathbf{V}^T) \cdot \mathbf{M}$, i.e., when I_0 is set to zero. From equation (7.31) it follows that for an observed Moran's I_0 the spectrum of eigenvalues of $\mathbf{M} \cdot [\frac{1}{2}(\mathbf{V} + \mathbf{V}^T) - I_0 \cdot \mathbf{I}] \cdot \mathbf{M}$ is translated by I_0 (i.e., $\{\lambda_1 - I_0, \ldots, \lambda_{n-k} - I_0, \underbrace{0, \ldots, 0}_{k \text{ zeros}}\}$). Due to the rank defect

of the projection matrix k eigenvalues have to remain zero. Consequently, while it is possible that due to inherent characteristics of the spatial link matrix $\mathbf{V}$ some of the eigenvalues $\{\lambda_1, \ldots, \lambda_{n-k}\}$ may be zero, these eigenvalues are still valid entries and equal I_0 for any observed Moran's $I_0 \neq 0$. The importance of these valid zero eigenvalues must be emphasized for local spatial link matrices which possess only two non-zero eigenvalues.

7.2.3   Bounding Distributions of Moran's I to Obtain Independence from the Projection Matrix

While with increasing computing power practical applications of bounding distributions of Moran's I are at least for the moment only of historical interest their properties give quite a few theoretical insights. Durbin and Watson (1950, see also the corrected reprint in Johnson and Kotz, 1992) have devised, under the assumption of independence the bounding distributions, to become independent from the regression matrix $\mathbf{X}$. This independence permitted them to work with a known spectrum of eigenvalues of their serial link matrix and to handle the regression residuals as if these were stochastically independent. The basic result is that there are two sequences of eigenvalues based only on the plain spatial link matrix $\frac{1}{2}(\mathbf{V} + \mathbf{V}^T)$ which bind the exact spectrum of eigenvalues of $\mathbf{M} \cdot \frac{1}{2}(\mathbf{V} + \mathbf{V}^T) \cdot \mathbf{M}$. In fact, these bounds follow from the Cournat-Fisher inequalities (see Courant and Hilbert, 1953, p 33). For a proof of these inequalities I refer here to the literature.

Historically, another strategy to obtain pseudo-independent regression residuals $\widetilde{\varepsilon}$ has caught much attention in the econometric literature in the seventies and eighties. These approaches are based on modified regression estimators which suppress an arbitrary set of k residuals due to the k constraints induced by the design matrix $\mathbf{X}$. Recall that ordinary least-squares generates regression residuals $\widehat{\varepsilon}$ which are orthogonal to the design matrix (i.e., $\mathbf{0} = \mathbf{X}^T \cdot \widehat{\varepsilon}$). These Best Linear Unbiased with Scalar co-variance matrix (BLUS) estimators, however, compute only a residual vector $\widetilde{\varepsilon}$ with $n - k$ component and leave the remaining k component undetermined (see Theil, 1965, and Koerts and Abrahamse 1969). While the choice on which k residuals should be suppressed is fairly simple for the time series situation (usual the first k residuals), this problem becomes considerably more complex in the cross-sectional situation: it is difficult to pick k boundary spatial objects in multi-directional spatial processes where the observations do not have any natural ordering. This has been noted *inter alia* by Cliff and Ord (1973), Bartels and Hordijk (1977), and Brandsma and Ketellapper (1979). Another problem with BLUS estimators is the reduction in the degrees

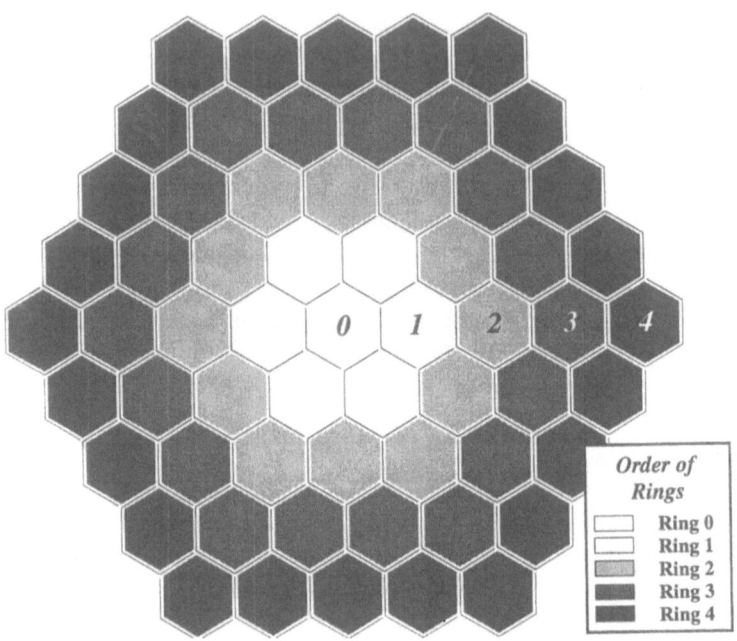

Fig. 7.1. Hexagonal spatial structure arranged in rings around a central cell

of freedom due to the k omitted observations. It has been noted in the literature that exact tests utilizing the full set of observation possess higher power than tests based on BLUS residuals (for a review see King, 1987).

Let the diagonal eigenvalue matrix of the plain spatial link matrix $\frac{1}{2}(\mathbf{V} + \mathbf{V}^T)$ be denoted by $\mathbf{\Gamma} \equiv \operatorname{diag}(\gamma_1, \ldots, \gamma_n)$ with $\gamma_i \leq \gamma_{i+1}$. Furthermore, the special situation that there might be s independent linear combinations of the variables in the design matrix $\mathbf{X}$ which are identical with some eigenvectors of the plain spatial link matrix needs to be considered. In this case the spectrum of eigenvalues $\{\gamma_1, \ldots, \gamma_n\}$ needs to be pruned by those associated eigenvalues so that the re-sequenced spectrum is given by $\{\gamma_1, \ldots, \gamma_{n-s}\}$. Then part (b) of the *Durbin and Watson Lemma* states that the relevant exact eigenvalues λ_i are bound from below and above by $\gamma_i \leq \lambda_i \leq \gamma_{k+i-s}$ with $i \in \{1, \ldots, n-k\}$. These bounds imply for the exact Moran's I that

$$I_L \equiv \frac{\sum_{i=1}^{n-k} \gamma_i \cdot \eta_i^2}{\sum_{i=1}^{n-k} \eta_i^2} \leq \frac{\sum_{i=1}^{n-k} \lambda_i \cdot \eta_i^2}{\sum_{i=1}^{n-k} \eta_i^2} \leq \frac{\sum_{i=1}^{n-k} \gamma_{k+i-s} \cdot \eta_i^2}{\sum_{i=1}^{n-k} \eta_i^2} \equiv I_U . \qquad (7.32)$$

Note that I_L and I_U are independent from the design matrix and depend solely on the eigenvalues of the spatial link matrix. Consequently, for an observed value I_0 of Moran's I the inequalities imply for its significances $\Pr(I_U \leq I_0) \leq$

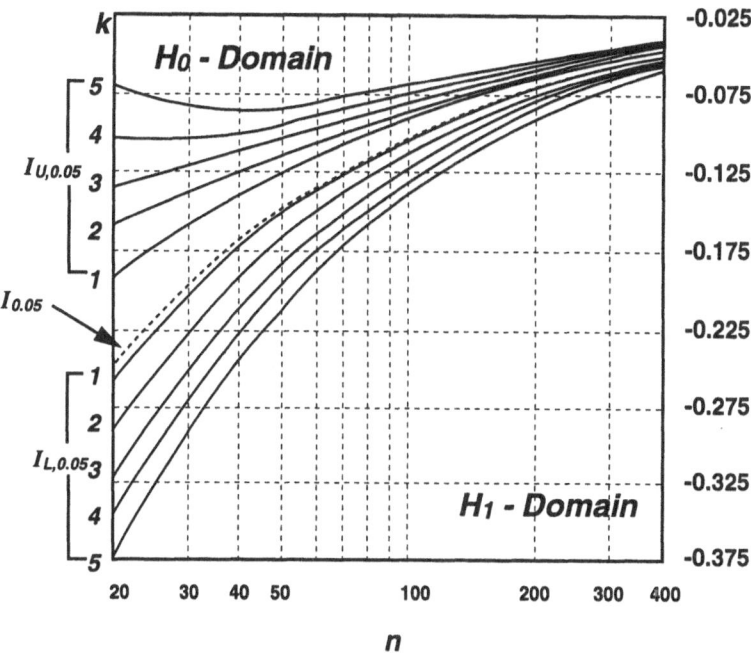

Fig. 7.2. Bounding 5% significance points ($k = 1, \ldots, 5$) of the upper and lower distribution of Moran's I for rings of hexagons

$\Pr(I \leq I_0) \leq \Pr(I_L \leq I_0)$. The lower bound is calculated based on the eigenvalues $\{\gamma_1, \ldots, \gamma_{n-k}\}$ and the upper bound is calculated based on the eigenvalues $\{\gamma_{k+1-s}, \ldots, \gamma_{n-s}\}$.

An Example To illustrate the behavior of the bounding distributions, in dependence of the number of spatial objects n and the number of omitted variables k, a tessellation based on rings of hexagons arranged in rings around a central cell is investigated. Figure 7.1 gives an example of these spatial structures with up to 4 rings and 61 cells in total. In general this sequence of hexagonal structures has $1 + 6 \cdot \sum_{i=1}^{r} i$ cells with $r = 1, 2, \ldots$ and r indicating the order of the most peripheral ring. Figure 7.2 gives the bounding distribution of a one-sided Moran's I test against negative spatial autocorrelation at a nominal significance level $\alpha = 0.05$ for 1 to 5 neglected independent variables including the constant vector. The spatial structure of the hexagons is specified in the C-coding scheme. Similar bounds can be constructed for a one-side test against positive spatial autocorrelation or a two-sided test against spatial autocorrelation of any sign.

For a given number of omitted independent variables, if the observed Moran's I_0 is larger than the upper bound, then the zero hypothesis can not be rejected whereas if I_0 falls below the lower bound it must be rejected at the nominal significance level $\alpha = 0.05$. However if it falls in-between both bounds the test is inconclusive and an exact test should be applied. As expected the inconclusive area becomes wider the more variables are neglected and it becomes narrower as the number of cells increases. Nevertheless, for a hexagonal structure with 61 cells and 5 neglected explanatory variables the range $|I_U - I_L|$ of the inclusive area is still larger than 0.1.

The Relation between Spatial Structures and the Design Matrix In Figure 7.2 the exact significance points of Moran's I for residuals defined solely as variation around the mean are shown by a dashed line. These exact significance points almost coincide with the lower bounding distribution of I_L for $k = 1$. It is known (see the Perron-Frobenius theorem 4.1 in Schwenk and Wilson, 1978) that the principal eigenvector $\mathbf{k}_n$ of the plain binary spatial link matrices consists solely of positive elements, thus any other eigenvector must have at least one negative element to be orthogonal to the principal eigenvector. In addition, the unity vector $\mathbf{1}$ modeling the variation around the mean is also positive for all its elements and thus the value of the direction-cosine between the principal eigenvector $\mathbf{k}_n$ and the unity vector $\mathbf{1}$ will be larger than for any other eigenvector of the binary spatial link matrix. Actually, for the rings of hexagons the unity vector $\mathbf{1}$ lies almost perfectly in the 1-dimensional subspace of the principal eigenvector $\mathbf{k}_n$. The vector of the direction-cosines between the eigenvectors of $\frac{1}{2}(\mathbf{V} + \mathbf{V}^T)$ and the unity vector $\mathbf{1}$ for a 7 cell hexagonal structure (one ring around a central cell) are $(0.00, 0.20, 0.00, 0.00, 0.00, 0.00, 0.98)^T$. For the unity vector $\mathbf{1}$ these values are actually the sums of the components of each eigenvector $\mathbf{k}_i$. The projection matrix $\mathbf{M}_{(1)} \equiv (\mathbf{I} - \mathbf{1} \cdot \mathbf{1}^T/n)$ defined solely on the unity vector $\mathbf{1}$ forces all the eigenvectors of $\mathbf{M}_{(1)} \cdot \frac{1}{2}(\mathbf{V} + \mathbf{V}^T) \cdot \mathbf{M}_{(1)}$ to be orthogonal to the unity vector (see de Jong et al., 1984), which in turn is almost collinear with the principle eigenvector $\mathbf{k}_n$ of $\frac{1}{2}(\mathbf{V} + \mathbf{V}^T)$. Thus, allowing for shifts, both spectra of eigenvalues are almost identical.

For example, the shifted spectrum $\{\lambda_i\}$ of $\mathbf{M}_{(1)} \cdot \frac{1}{2}(\mathbf{V} + \mathbf{V}^T) \cdot \mathbf{M}_{(1)}$ and the original spectrum $\{\gamma_i\}$ of $\frac{1}{2}(\mathbf{V} + \mathbf{V}^T)$ for a 7 cell hexagonal structure are shown in Table 7.1. The eigenvalues in italics are both excluded from calculating the distributions of the exact Moran's I and the bounded I_L due to Durbin and Watson's Lemma. The high degree of correspondence in both reduced sets of eigenvalues forces the distributions of both Moran's I and Moran's I_L to be almost identical. Note that only those eigenvalues, for which the direction cosine is not zero, their values change and that the projection matrix $\mathbf{M}_{(1)}$ reduces the rank by exactly $k = 1$. For the exact relationship between the eigenvalues of both spectra when the regression matrix $\mathbf{X}$ consists of only one variable see Durbin and Watson (1950, p 414).

A similar line of argument has been used by Durbin and Watson (1971) to become independent of $\mathbf{X}$. They proposed an approximation to the exact distribution of their d-statistic based on the upper bounding distribution. Those eigenvectors of a serial link matrix used by Durbin and Watson which correspond to the smallest eigenvalues exhibit a low rate of change between successive

Table 7.1. Relationship between γ_i and λ_i for a projection matrix defined solely on the unity vector. The values in italics must be excluded from the calculation of the distributions I_L and I

i	γ_i	Shift	λ_i
1	−0.5833	→	−0.5833
2	−0.4800	→	−0.4167
3	−0.2917	→	0.2917
4	−0.2917	→	−0.2917
5	0.2917	↘	*0.0000*
6	0.2917	↘	0.2917
7	*1.0633*		0.2917

elements. See Durbin and Watson (1971, p 1) for exact formulas of the eigen-vectors and -values of their serial link matrix, in particular the eigenvector $\mathbf{k}_1$ associated with the smallest eigenvalue is constant. Furthermore, variables in economic time series normally show a low rate of change between consecutive time periods. Therefore, the subspace spanned by these eigenvectors is close to the subspace spanned by the independent time series variables. Consequently, the shifted eigenvalue spectrum due to the projection matrix resembles that of the largest eigenvalues of Durbin and Watson's serial link matrix.

Discussion Bounding significance points can be calculated by adjusting for partial knowledge of the design matrix $\mathbf{X}$. Clearly, explicitly accounting for a known subset of regressors reduces the inconclusive region of the bounds test procedure. With this in mind, King (1981) showed for a partitioned design matrix $\mathbf{X} = [\mathbf{X}_1 \mid \mathbf{X}_2]$ that part (b) of the Durbin and Watson Lemma can be extended. He showed that again the re-sequenced spectrum of eigenvalues based on the full projection matrix $\mathbf{M} = \mathbf{I} - \mathbf{X}(\mathbf{X}^T\mathbf{X})^{-1}\mathbf{X}^T$ is interlaced into the spectrum of eigenvalues based on the partial design matrix $\mathbf{M}_1 = \mathbf{I} - \mathbf{X}_1(\mathbf{X}_1^T\mathbf{X}_1)^{-1}\mathbf{X}_1^T$. This allowed him to account for seasonal dummy variables in a serial regression model and thus attained more precise bounds for Durbin and Watson's statistic on serial autocorrelation.

Under the alternative hypothesis bounding distributions of the power function are unknown. However for serial autocorrelated processes with a special co-variance structure, Tillman (1975) found an upper bound and an attainable lower bound of the exact spectrum of eigenvalues.

Tests on spatial autocorrelation within the maximum likelihood domain (see Chapter 4.3.1 for the likelihood ratio test and the Lagrangian multiplier test) ignore the impact of the projection matrix because they are embedded in an asymptotic context. While it is true that asymptotically a projection matrix $\mathbf{M}$ approaches for a given number k of explanatory variables the neutral identity matrix $\mathbf{I}$ (see Upton and Fingleton, 1985, p 337) the inconclusive region in Figure 7.2 even for $n = 397$ demonstrates very well that in the spatial context maximum likelihood estimators need an abundant amount of observations to become feasible.

Due to the fixed structure of Durbin and Watson's serial link matrix, bounds of their statistic have been tabulated quite extensively under varying conditions. However, in the spatial situation the specification of the link matrix varies from spatial setting to spatial setting. Furthermore, generalized spatial link matrices are at the discretion of the analyst and she/he can choose among several coding schemes. Therefore, only within the narrow realm of a specific research project where the data are screened in an initial analysis for spatial autocorrelation, a tabulation of bounding significance points can make sense. Under such an agenda the spatial setting remains fixed whereas several combinations of explanatory variables are probed for their appropriateness and the residuals of such models are tested for spatial autocorrelation. While the calculation of the autocorrelation test statistics like Moran's *I* requires a full knowledge of the regression model, using the bounding significance points as benchmarks requires no (or only partial) knowledge of the regression model and are therefore swiftly performed. However, for medium sized spatial models (i.e., a few hundred spatial objects) computing time is not that crucial anymore, so that a bunch of exact autocorrelation tests can be directly applied.

Another advantage of the bounding distributions in the case of nonlinear regression analysis is actually their independence from the regression matrix $\mathbf{X}$ and therefore from the projection matrix $\mathbf{M}$. Normally, nonlinear regression procedures do not give us an equivalent to the projection matrix $\mathbf{M}$. Due to the absence of this matrix, the exact distribution of Moran's *I* cannot be calculated but, nevertheless, its significance can be assessed by the bounding distributions. For instance, in Poisson, loglinear or logistic regression the so-called adjusted residuals are asymptotically normally distributed (see Agresti, 1990, p 433). Based on these adjusted residuals, the observed Moran's I_0 can be calculated and tested for spatial independence by means of the bounding distributions.

In my assessment, the most important application of the bounding distributions is for autocorrelation tests of the regression residuals with lagged dependent variables. Durbin (1970) observed that under this situation the appropriate critical value of the Durbin and Watson statistic for serial autocorrelation can not be computed. King and Wu (1991) have devised a test procedure and shown its statistical feasibility for a dynamic regression model $y_t = \alpha_1 \cdot y_{t-1} + \cdots + \alpha_p \cdot y_{t-p} + \mathbf{x}_t^T \cdot \boldsymbol{\beta} + \varepsilon_t$ which uses the p-th bounding distributions. Note that the disturbance ε_t is contemporarily independent from the lagged dependent variables $\mathbf{y}_{t-1}, \ldots, \mathbf{y}_{t-p}$. For the $t - p$ observations the full design matrix $\mathbf{Z}$ can be written in block form as $\mathbf{Z} = [\mathbf{y}_{t-1}, \ldots, \mathbf{y}_{t-p} \mid \mathbf{X}]$ and the p-th bounding distributions are calculated only by use of the projection matrix $\mathbf{M}_X = \mathbf{I} - \mathbf{X}(\mathbf{X}^T\mathbf{X})^{-1}\mathbf{X}^T$. Unfortunately, this procedure does not generalize easily for spatial models: spatially lagged dependent variables are correlated with the disturbances at the reference locations due to the multi-directional nature of spatial processes. Clearly more research on the feasibility of King and Wu's procedure is needed in the spatial situation.

Chapter 8

The Shape of Moran's *I* Exact Distribution

As already pointed out earlier the spectrum of eigenvalues is of vital importance for the derivation of the exact distribution of Moran's *I* because it determines the shape of the distribution function. This chapter approaches the shape of the exact distribution from two different ends. First, the feasible range of Moran's *I* will be derived which is bound and identical under the assumption of spatial independence as well as under the assumption of the presence of the significant spatial process. Then the distribution of the eigenvalues within the feasible range and its impact on the shape of the exact distribution will be sketched for small and large sample sizes. This permits me to speculate under which circumstances we are allowed to approximate the exact distribution by a normal distribution.

8.1 The Feasible Range of Moran's *I*

The feasible range of Moran's *I* has been derived under the zero hypothesis by de Jong et al. (1984) where the co-variance matrix $\mathbf{\Omega}$ is diagonal, i.e., $\mathbf{\Omega} \equiv \sigma^2 \cdot \mathbf{I}$, and can therefore be neglected. While de Jong et al. (1984) concentrate only on a projection matrix solely defined on the constant unity vector $\mathbf{1}$ their results are easily generalized for arbitrary projection matrices $\mathbf{M}$. The feasible range of Moran's *I* is given by the smallest and largest eigenvalue of $\mathbf{M} \cdot \frac{1}{2}(\mathbf{V} + \mathbf{V}^T) \cdot \mathbf{M}$ with $\lambda_1 < 0 < \lambda_n$. Disregarding for a moment the impact of the projection matrix $\mathbf{M}$, the first eigenvalue must be smaller than zero and the n-th eigenvalue must be greater than zero because the elements on the diagonal of $\frac{1}{2}(\mathbf{V} + \mathbf{V}^T)$ are all zero by definition and, therefore, the relation $\operatorname{tr}\left(\frac{1}{2}(\mathbf{V} + \mathbf{V}^T)\right) = \sum_{i=1}^{n} \lambda_i = 0$ between the diagonal elements of a spatial link matrix and its eigenvalues must hold.

The following derivation will be more general than the one given by de Jong et al. For a matrix $\mathbf{B}$ with full rank it is known that general ratios of quadratic forms

$$\frac{\delta^T \cdot \mathbf{A} \cdot \delta}{\delta^T \cdot \mathbf{B} \cdot \delta} \tag{8.1}$$

in i.i.d. normal distributed variables δ_i are bound by the minimum and maximum eigenvalues of $\mathbf{B}^{-1} \cdot \mathbf{A}$ (see Jones, 1987, p 129). Since in our case the denominator $\mathbf{B}$ is singular due to the rank defect of the projection matrix we need to apply the generalized inverse. Three properties of this *g*-inverse are important here:

Theorem 7 (Dhrymes (1984, p 87)). *If $\mathbf{Q}$ is quadratic and non-singular, then*

$$\mathbf{Q}^{-1} = \mathbf{Q}_g \ .$$

Theorem 8 (Dhrymes (1984, p 87)). *If $\mathbf{Q}$ is symmetric and idempotent, then*

$$\mathbf{Q} = \mathbf{Q}_g \ .$$

Theorem 9 (Dhrymes (1984, p 87 and p 88)). *If $\mathbf{Q} = \mathbf{P}^T$, then*

$$(\mathbf{Q} \cdot \mathbf{P})_g = \mathbf{P}_g \cdot \mathbf{Q}_g \ .$$

By making use of these theorems and by noting that the denominator in equation (7.6) is symmetric and that $\mathbf{M}$ is idempotent[1] we can write

$$\frac{\mathbf{\Omega}^{T\frac{1}{2}} \cdot \mathbf{M} \cdot \frac{1}{2} \cdot (\mathbf{V} + \mathbf{V}^T) \cdot \mathbf{M} \cdot \mathbf{\Omega}^{\frac{1}{2}}}{\mathbf{\Omega}^{T\frac{1}{2}} \cdot \mathbf{M} \cdot \mathbf{\Omega}^{\frac{1}{2}}} \tag{8.2}$$

$$\Leftrightarrow \mathbf{\Omega}^{-\frac{1}{2}} \cdot \underbrace{\mathbf{M}_g}_{\mathbf{M}} \cdot \underbrace{\mathbf{\Omega}^{T-\frac{1}{2}} \cdot \mathbf{\Omega}^{T\frac{1}{2}}}_{\mathbf{I}} \cdot \mathbf{M} \cdot \frac{1}{2}(\mathbf{V} + \mathbf{V}^T) \cdot \mathbf{M} \cdot \mathbf{\Omega}^{\frac{1}{2}} \tag{8.3}$$

$$\Leftrightarrow \mathbf{\Omega}^{-\frac{1}{2}} \cdot \mathbf{M} \cdot \frac{1}{2}(\mathbf{V} + \mathbf{V}^T) \cdot \mathbf{M} \cdot \mathbf{\Omega}^{\frac{1}{2}} \tag{8.4}$$

$$\Rightarrow \mathbf{M} \cdot \frac{1}{2}(\mathbf{V} + \mathbf{V}^T) \cdot \mathbf{M} \cdot \mathbf{\Omega}^{\frac{1}{2}} \cdot \mathbf{\Omega}^{-\frac{1}{2}} \tag{8.5}$$

$$\Leftrightarrow \mathbf{M} \cdot \frac{1}{2}(\mathbf{V} + \mathbf{V}^T) \cdot \mathbf{M} \tag{8.6}$$

The change in sequence of matrices in equation (8.5) is allowed because it is known (Macduffee, 1964, Theorem 16.2) that the spectrum of eigenvalues of two matrices does not change by reversing the order of multiplication of these two matrices.[2] Therefore, the bounds of Moran's I under the zero hypothesis and under alternative hypotheses are equivalent. Consequently, the probability of observing a Moran's I outside the range $[\lambda_1, \lambda_n]$ is zero (i.e., $\Pr(I \leq \lambda_1) = \Pr(I \geq \lambda_n) = 0$).

8.2   Specific Eigenvalue Spectra and Moran's I's Distribution

The previous section has given the feasible range of Moran's I in terms of the smallest and largest eigenvalues and Section 6.1 has shown that we can neglect

---

[1] Recall that idempotent matrices are characterized by $\mathbf{M} = \mathbf{M} \cdot \mathbf{M}$.
[2] Suppose that λ is a root of the product of the two matrices $\mathbf{A}$ and $\mathbf{B}$. Then there exists a non-zero characteristic vector $\mathbf{p}$ such that $\mathbf{A} \cdot \mathbf{B} \cdot \mathbf{p} = \lambda \cdot \mathbf{p}$. Pre-multiplying this equation by $\mathbf{B}$ gives $\mathbf{B} \cdot \mathbf{A} \cdot (\mathbf{B} \cdot \mathbf{p}) = \lambda \cdot (\mathbf{B} \cdot \mathbf{p})$. Therefore, the eigenvalues of $\mathbf{A} \cdot \mathbf{B}$ and $\mathbf{B} \cdot \mathbf{A}$ are both the same, however, their characteristic vectors are different with $\mathbf{p}$ and $(\mathbf{B} \cdot \mathbf{p})$, respectively.

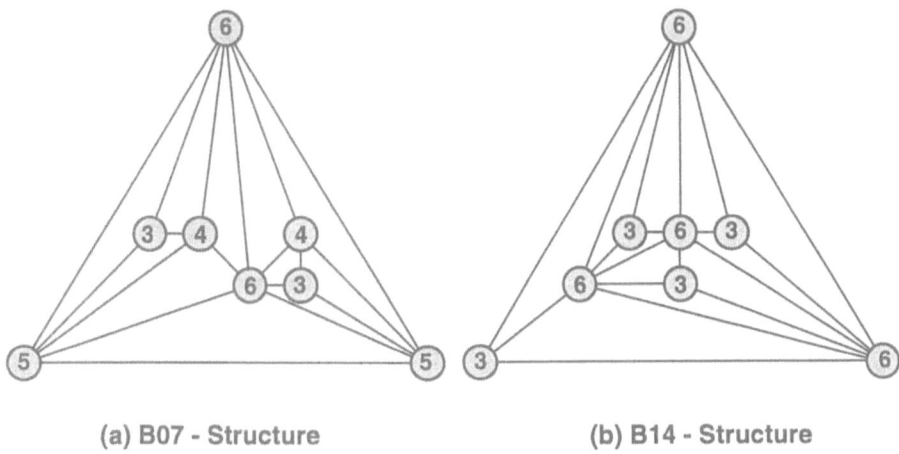

(a) B07 - Structure   (b) B14 - Structure

Fig. 8.1. Triangulation and object linkage degrees of (a) the $B7$ structure and (b) the $B14$ structure

all eigenvalues in the calculation of an exact significance level of Moran's I_0 which equal zero. To show the impact of the spectrum of eigenvalues on the distribution of Moran's I we have to restrict our attention to the assumption of spatial independence. Only under this condition the re-sequenced spectrum of the $n-k$ eigenvalues is up to a translation by I_0 fixed (see Chapter 7.2.2).

8.2.1   Small Irregular Spatial Tessellations

First the complete distribution of Moran's I for two different spatial structures is given. The underlying spectra of eigenvalues were calculated from their spatial link matrices in the C-coding scheme. These two spatial structures belong to an exhaustive set of fourteen maximally connected planar spatial structures called the B-series with a fixed number $n = 8$ nodes and an overall connectivity $D = 36$. Any additional edge would force these graphs to become non-planar (i.e., some edges would cross each other). These structures were generated by an exhaustive enumeration procedure outlined in Boots and Royle (1991). This small number of spatial objects has been chosen so that the deviance of the exact distribution from the normal approximation will be visually noticeable. Figure 8.1 displays two members of the B-series (i.e., the $B7$ and the $B14$ structure) which are with respect to the distribution of the local linkage degree vector extreme. The $B7$ structure has a degree vector $\mathbf{d}_{(B7)} = (3, 3, 4, 4, 5, 5, 6, 6)^T$ which is evenly distributed whereas the degree vector $\mathbf{d}_{(B14)} = (3, 3, 3, 3, 6, 6, 6, 6)^T$ of $B14$ structure is extremely bi-modal distributed.

Figure 8.2 displays the distribution of Moran's I for both structures based on the projection matrix $\mathbf{M}_{(1)}$ which models the variation of observed data values around their mean. The density functions $\widetilde{f}(I_s) \equiv [F(I_{s+1}) - F(I_{s-1})]/2$

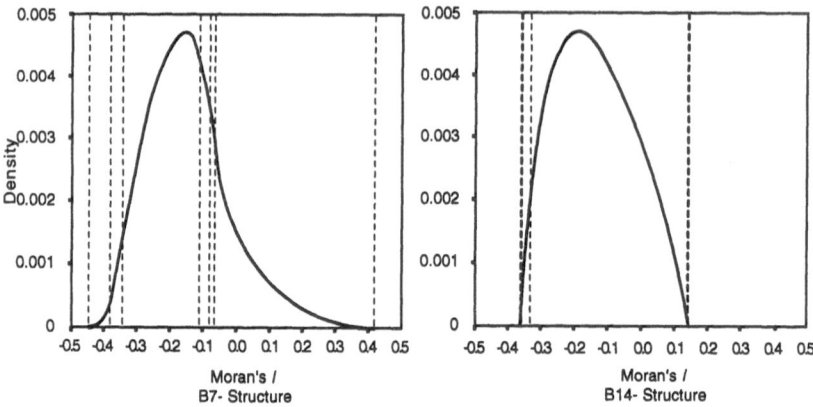

Fig. 8.2. Density function of Moran's I for the $B7$- and the $B14$-structure with the underlying resequenced spectra of eigenvalues superimposed on top of it

are derived from the exact distribution functions $F(I)$ by taking the simple first differences at the significance points I_{s+1} and I_{s-1}. On average the density functions rest on 600 equidistant supporting points n_s along the axis of the feasible range of Moran's I. The accuracy for this number of supporting points is sufficiently precise which can be checked by comparing the theoretical moments of Moran's I (see Chapter 9 for their derivation) against the approximated central moments based on the density function (i.e., $m_p(I) = \frac{1}{n_s} \sum_{s=1}^{n_s} \widetilde{f}(I_s) \cdot (I_s - \mathrm{E}[I])^p$ with $\mathrm{E}[I] = \frac{1}{n_s} \sum_{s=1}^{n_s} \widetilde{f}(I_s) \cdot I_s$, $p = 2, 3, \dots$ and $I_s \in [\lambda_1, \lambda_{n-k}]$). The first four moments for both modes of calculation are identically up to the fourth significant digit.

For both spatial structures we observe as claimed by theory that the distribution of Moran's I varies within its bounds $[\lambda_1, \lambda_{n-k}]$. While for the $B7$-structure the distribution of the remaining 7 re-sequenced eigenvalues is more disperse for the $B14$-structure it is clustered at the lower and higher bound. In fact, for the $B14$-structure 3 eigenvalues at either end are identical and have been visually slightly jittered in Figure 8.2 to indicate their multitude. The one extreme positive eigenvalue for the $B7$-structure forces the distribution of Moran's I to have a long positive tail area so that it is positively skewed. Contrary, the shape of the distribution of Moran's I for the $B14$-structure is highly compact due to the two clusters at the lower and higher bound. Therefore, for the $B14$-structure the kurtosis of the Moran's I's distribution is low.

These two examples indicate that the distribution of eigenvalues exercises sort of a pulling-in force on the shape of the density function of Moran's I. The proposed relation between the spectrum of eigenvalues and the distribution of Moran's I allows me to speculate that a spectrum of eigenvalues with a large cluster of values in the center and a few eigenvalues at both bounds will result in a distribution of Moran's I with a large kurtosis. We will see below and in the discussion of local Moran's I_i (see Chapter 11.2.2) that this speculation is viable.

8.2.2 Large Regular Spatial Tessellations

Previous discussion has shown the dependence of Moran's I's distribution on the distribution of the eigenvalues for small underlying spatial tessellations. This section will exemplify that for well-behaved spatial structures the spectrum of eigenvalues becomes with increasing n more disperse. Therefore, the distribution of Moran's I approaches the normal distribution.

Four different regular spatial structures (i.e., a triangular tessellation arranged in a triangle and in a rectangle, a rectangular square lattice, and a hexagonal tessellation arranged in a square) are investigated for an increasing sequence of 64, 256, and 1024 spatial objects. Figure 3.2 gave an example of these spatial tessellations for $n = 256$. Because the size of the tessellations increases in quadruples it does not matter for the triangular and square lattices if we argue in terms of the in-fill asymptotics or in terms of the increasing-domain asymptotics (see Cressie (1991) for a general discussion on the asymptotic perspectives in spatial statistics). For these two regular structures we can consider an increasing number of spatial objects as a systematic subdivision of the existing spatial objects (in-fill aysmptotics) or as spatial objects newly appended around an existing tessellation (increasing-domain asymptotics). Investigations of rectangular spatial structures appeal to remote sensing due to the perpendicular arrangement of the receptors, however, they have only limited value in human geography. Hexagonal spatial structures are much more coherent with irregular planar tessellations observed in empirical work (see Chapter 3.2.1). I feel therefore uncomfortable with simulation experiments and theoretical discussions conducted on rectangular spatial structures in spatial statistics and spatial econometrics. As will be shown below the hexagonal spatial tessellation differs substantially from the triangular and rectangular tessellations.

The Distribution of Eigenvalues for Regular Spatial Structures Depending on the spatial structure the distributions of their related eigenvalues show marked differences, however, no more isolated clusters of eigenvalues. Figure 8.3 shows these distributions for the triangular, rectangular, and hexagonal structures with 1024 spatial objects. Up to the marginal impact of different geometrical symmetry properties, virtually no difference can be observed in dependence of the edge structure of the two triangular tessellations. Note that average linkage degree in both arrangements is identical. The distribution of their eigenvalues is bi-modal. This indicates as for the $B14$-structure that the kurtosis of the distribution of Moran's I related to triangular spatial tessellations must be relatively small. In contrast, the spectrum of eigenvalues for rectangular spatial tessellations exhibits clustering in its center with heavy pulling tail eigenvalues. One therefore can expect a relatively large kurtosis of the distribution of Moran's I. Finally, for hexagonal tessellations the spectrum of eigenvalues is

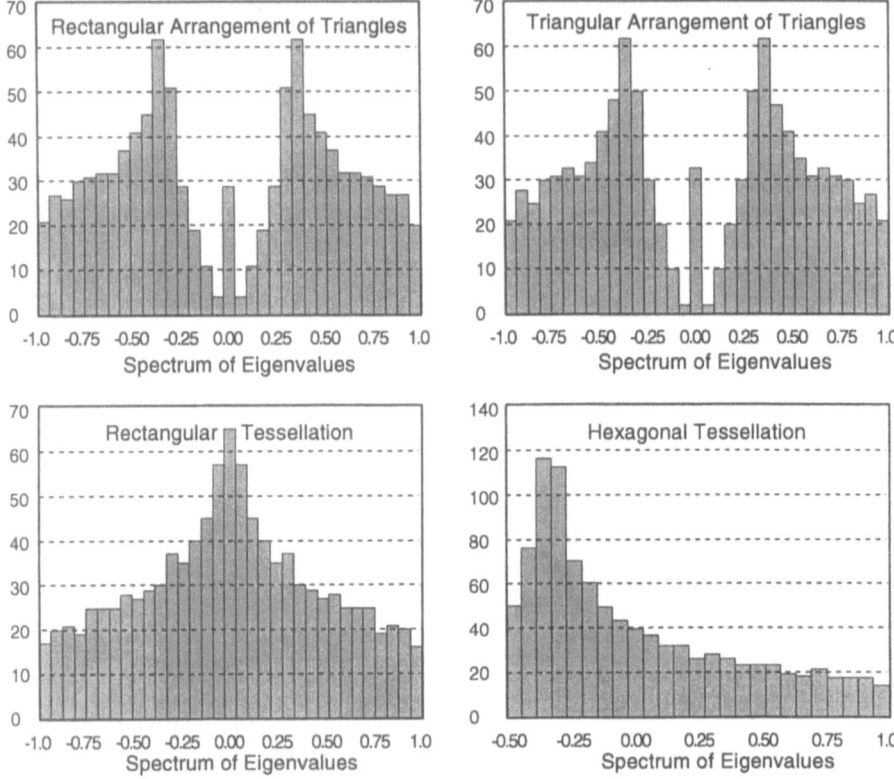

Fig. 8.3. Distribution of the spectra of eigenvalues depending on different regular spatial tessellations for 1024 spatial objects

highly skewed and the bounds are not symmetric around zero. This shape of the spectrum of eigenvalues with bounds roughly $[-0.5, 1.0]$ is also typical for most irregular planar tessellations. The long positive tail of eigenvalues forces the distribution of Moran's I to be relatively positive skewed.

The asymmetry in the spectrum of eigenvalues for hexagonal tessellations and a negative bound only half the magnitude of the positive bound can be best understood by studying a checkerboard pattern of alternating black and white values distributed over a hexagonal tessellation. Figure 8.4 indicates that it is not possible to distribute black and white values over the map in such a manner that the neither black nor white values share a similar neighbor. Therefore, in the interior of this tessellation a white cell has three black and three white neighbors neutralizing each other's autocorrelation potential. However, a black cell has six white neighbors leading to a maximal negative local autocorrelation level. While it is possible to cluster black and white values in this tessellation in such a way that it leads to maximal positive global autocorrelation, it is impossible to mix

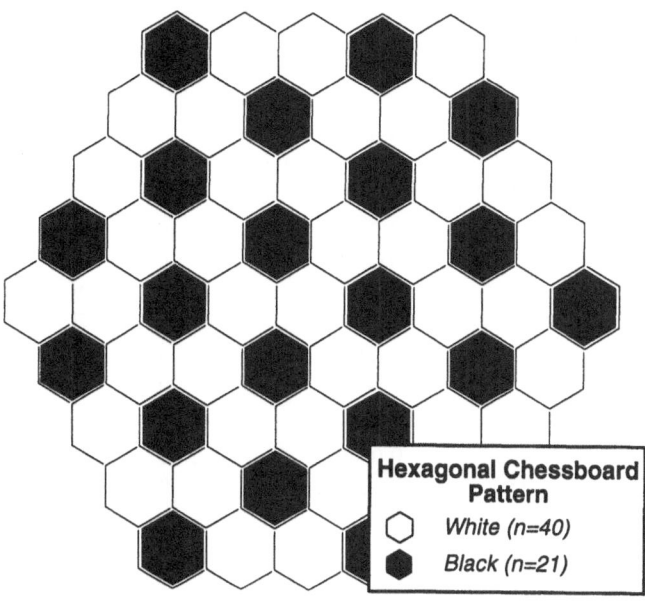

Fig. 8.4. Alternating black and white pattern in a hexagonal tessellation indicating maximal attainable negative autocorrelation

the black and white values in such a manner that each cell contributes maximally to negative global autocorrelation. This restriction leads to a reduced range of the negative autocorrelation level which is also observable in empirical irregular tessellations. In contrast, rectangular and triangular tessellations do not possess this asymmetry in their spectra of eigenvalues because black and white values can be distributed in the Rook-specification alternating over these tessellations without contacts to similar cells.

Validity of the Normal Approximation for Large Regular Spatial Structures To show the relation between a spectrum of eigenvalues and the distribution of Moran's I Table 8.1 gives its moments defined on the projection matrix $M_{(1)}$ in the C-coding scheme. The normal distribution sets the benchmarks for comparing the figures in this table: the skewness of the normal distribution is 0 which means it is symmetric and its kurtosis is 3. As suspected the kurtosis for triangular tessellations is below 3 whereas the kurtosis of rectangular tessellations is greater than 3. Furthermore, the skewness of hexagonal tessellations is, as assumed, positive.

Table 8.1. Moments of Moran's I and average linkage degrees $\frac{D}{n}$ in dependences of the spatial tessellations and their size n

Tessellation	n	$\frac{D}{n}$	Expectation	Variance	Skewness	Kurtosis
Triangles	64	2.625	−0.015873	0.011369	−0.000158	2.929960
(Rectangular	256	2.813	−0.003922	0.002746	−0.000008	2.985230
Arrangement)	1024	2.906	−0.000978	0.000670	−0.000000	2.996209
Triangles	64	2.625	−0.015873	0.011365	−0.000340	2.944440
(Triangular	256	2.813	−0.003922	0.002746	−0.000012	2.985320
Arrangement)	1024	2.906	−0.000978	0.000670	−0.000000	2.996215
	64	3.500	−0.015873	0.008405	−0.011953	3.038531
Rectangles	256	3.750	−0.003922	0.002052	−0.001414	3.011604
	1024	3.875	−0.000978	0.000502	−0.000174	3.002970
	64	5.032	−0.015873	0.005665	0.270432	3.081494
Hexagons	256	5.508	−0.003922	0.001386	0.142315	3.023869
	1024	5.752	−0.000978	0.000338	0.071931	3.006016

In order to show that for these three regular tessellations the distribution of Moran's I approaches the normal distribution a theorem due to Hoeffding (1952) can be used:

Theorem 10. *Let $F(y)$ be a distribution function uniquely determined by its theoretical moments, μ_p, and let $\{F_n(y)\}$, $n = 1, 2, \ldots$, be a sequence of distribution functions with moments μ_{np}. If $\mu_{np} \to \mu_p$ as $n \to \infty$ for $p = 1, 2, \ldots$, then $F_n(y) \to F(y)$ in probability at every continuity point of $F(y)$.*

Without following a formal discussion we observe from Table 8.1 that as n increases the skewness and kurtosis of Moran's I approach those of the normal distribution. Furthermore, the variance of Moran's I is inversely proportional to the size of the regular tessellations. Note, the small difference in the skewness

Table 8.2. Convergence of Moran's I to the normal distribution in dependence of different spatial tessellations and the sample size

Tessellation	n	$\int \lvert \Phi(I) - F(I) \rvert \cdot dI$	$\max \lvert \Phi(I) - F(I) \rvert$
Triangles	64	0.000405	0.001405
(Rectangular	256	0.000050	0.000346
Arrangement)	1024	0.000006	0.000087
	64	0.000286	0.001393
Rectangles	256	0.000035	0.000312
	1024	0.000004	0.000072
	64	0.003333	0.018293
Hexagons	256	0.000853	0.009501
	1024	0.000214	0.004788

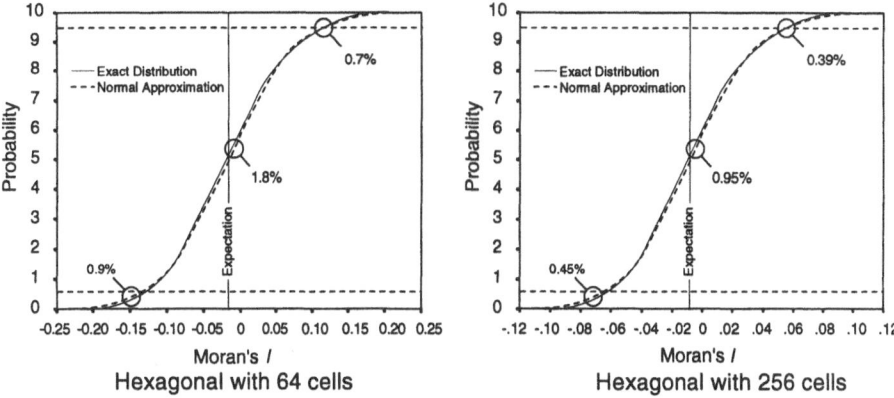

Fig. 8.5. Comparison of Moran's I's cumulative distribution function for hexagons with the normal distribution under the influence of an increasing number of spatial objects

and kurtosis between the triangles arranged either rectangularly or triangularly. Consequently, the distribution of Moran's I resembles that of the normal distribution for these spatial tessellations and large n. However, this convergence in distribution is by order 2 slower for Moran's I defined on a hexagonal tessellation than for the other two tessellations. Also the discrepancy between the exact distribution of Moran's I and the normal approximation is much larger for any tessellation size. Table 8.2 underlines this behavior where $\Phi(I)$ denotes the cumulative normal distribution and $F(I)$ the exact distribution of Moran's I.

Figure 8.5 visualizes the differences in the cumulative distribution functions for the hexagons defined either on 64 or 256 spatial objects with that of the normal distribution. For hexagons defined on 1024 spatial objects a difference is virtually no longer visible. Note that the largest difference is always observed around the expectation of Moran's I where it matters least for statistical tests. However, in the tails of the distribution we observe another peak in the differences. Practically speaking, in confirmatory statistical analysis we need over 200 spatial observations to work with the normal approximation. In contrast, in exploratory statistical analysis around 100 observations are sufficient.

A formal discussion on the necessary and sufficient conditions for this asymptotic normality is given in Cliff and Ord (1981, pp 47–51) in terms of the cumulants $\kappa_p(I)$ of the numerator of Moran's I for $p \geq 3$ as

$$\frac{\kappa_p(I)}{[\kappa_2(I)]^{\frac{p}{2}}} \text{ be } o(1) \tag{8.7}$$

(i.e., the ratio of cumulants goes to zero for large n). The denominator in equation (8.7) relates to the variance and simply standardizes the numerator. For a definition of the cumulants in terms of the spectrum of eigenvalues see Chapter 9.1.2. Note however, that erroneously Cliff and Ord's cumulants have been defined in terms of n and not in terms of $n - k$ (this difference does not matter for large n). Other conditions can be found in Cliff and Ord (1973) and Sen (1976), however, these are only sufficient conditions.

The presence of the forces of a significant spatial process will tamper this convergence in probability to the normal distribution. Again without being formal, the expectation of Moran's I shifts towards either related bound of its feasible range (see Chapter 11.1 for details). This shift induces a large skewness in the distribution of Moran's I because its bounds are invariant under the alternative hypothesis.

Chapter 9

The Moments of Moran's I

Previous chapters have referred already a few times to the moments of the distribution of Moran's I. This chapter will derive them in general terms based on a ratio of quadratic forms in normally distributed random variables. Under the independence assumption the moments of Moran's I can be given in a closed form in terms of the spectrum of eigenvalues. This simplifies their calculation substantially. Equivalently, the moments can also be given in terms of the trace of $\mathbf{M} \cdot \frac{1}{2}(\mathbf{V} + \mathbf{V}^T) \cdot \mathbf{M}$. Under the forces of a significant spatial process we have to retreat again to the more demanding numerical integration procedures to attain the moments of Moran's I. In contrast to the calculation of the distribution function of ratios of quadratic forms which is well-known in statistics and econometrics, calculation of the conditional moments has been considered to a much lesser degree. By using simple statistical theory also the correlation between different Moran's I's and therefore between different spatial structures can be given under the assumption of spatial independence and conditionally on the forces of a global spatial process.

The moments derived here are not random estimators as they would be in simulation studies but fixed parameters of the distribution of Moran's I. Therefore, we can use the theoretical moments of Moran's I to approximate its distribution function by the method of moment estimators for well-known distribution functions (Ord and Stuart, 1994, p 225). Another use of the moments is to link an observed Moran's I_0 to the autocorrelation parameter ρ of a hypothetical Gaussian spatial process. These two applications of the moments of Moran's I will be discussed in more detailed here.

9.1 The Moments under the Assumption of Spatial Independence

To derive the moments of Moran's I under the assumption of spatial independence use will be made of a theorem due to Pitman (1937). This theorem allows one to derive the moments of numerator and denominator of Moran's I independently in the next subsection. Again the importance of the eigenvalue spectrum will be noticeable.

9.1.1 The Theorem by Pitman

Recall that Moran's I can be expressed in terms of the resequenced spectrum of eigenvalues $\{\lambda_1, \ldots, \lambda_{n-k}, \underbrace{0, \ldots, 0}_{k \text{ zeros}}\}$ of $\mathbf{M} \cdot \frac{1}{2}(\mathbf{V} + \mathbf{V}^T) \cdot \mathbf{M}$ by a ratio of the quadratic forms in independent normal variables $\eta_i \sim \mathcal{N}(0, \sigma^2)$, i.e.,

$$I(\eta) = \frac{u(\eta)}{v(\eta)} = \frac{\sum_{i=1}^{n-k} \lambda_i \cdot \eta_i^2}{\sum_{i=1}^{n-k} \eta_i^2} \ . \tag{9.1}$$

Now the theorem by Pitman (1937, see also Stuart and Ord, 1994, p 529 for a proof) states

Theorem 11. *If $X_1, X_2, \ldots, X_n$ are independent identically distributed normal variates, say $\mathcal{N}(0, 1)$, then any scale-free function of them, $h(X_1, X_2, \ldots, X_n)$ is distributed independently of $Q = \sum_{i=1}^{n} X_i^2$.*

Using Pitman's theorem the denominator $v(\eta)$, which is a sum of squares, and $I(\eta) \equiv \frac{u(\eta)}{v(\eta)}$, which is a scale-free function, are independent. Thus, we can write

$$\mathrm{E}\big[(I(\eta) \cdot v(\eta))^p\big] = \mathrm{E}[I^p(\eta)] \cdot \mathrm{E}[v^p(\eta)] \ , \quad p = 1, 2, \ldots \tag{9.2}$$

so that, by noting that $\mathrm{E}\big[(\frac{u(\eta)}{v(\eta)} \cdot v(\eta))^p\big] = \mathrm{E}[u^p(\eta)]$, we have

$$\mathrm{E}[I^p(\eta)] = \frac{\mathrm{E}[u^p(\eta)]}{\mathrm{E}[v^p(\eta)]} \ . \tag{9.3}$$

Consequently, the crude moments of Moran's I can be expressed as a ratio of the moments of its numerator and denominator.

9.1.2   The Expectations of Moran's I Numerator and Denominator

The denominator v of Moran's I is a sum of $n - k$ independent central χ^2-distributed variables each with one degree of freedom. Therefore (see Stuart and Ord, 1994), the denominator has the crude moments

$$\mathrm{E}[v^p] = \prod_{q=1}^{p} n - k + 2 \cdot (q - 1) \tag{9.4}$$

Also the numerator u is a sum of $n - k$ independent $\lambda_i \cdot \eta_i^2$ variables, where η_i^2 is again a central χ^2-distributed variable with one degree of freedom. The coefficients λ_i belong to the reduced spectrum of eigenvalues $\{\lambda_1, \ldots, \lambda_{n-k}\}$ of $\mathbf{M} \cdot \frac{1}{2}(\mathbf{V} + \mathbf{V}^T) \cdot \mathbf{M}$ where the k necessarily zero eigenvalues have been dropped. Hence (see Stuart and Ord, 1994, Eq. 15.56), the j-th cumulant of u is

$$\kappa_p(u) = 2^{p-1}(p - 1)! \cdot \sum_{i=1}^{n-k} \lambda_i^p \ . \tag{9.5}$$

In particular, we can express the expectation $\mathrm{E}[u]$ by the first cumulant, i.e., $\mathrm{E}[u] = \kappa_1(u) = \sum_{i=1}^{n-k} \lambda_i$. Therefore,

$$\mathrm{E}(I) = \frac{\sum_{i=1}^{n-k} \lambda_i}{n - k} = \overline{\lambda} \tag{9.6}$$

so that the variation of Moran's I around its mean can be expressed by using equation (9.1), i.e.,

$$I - \mathrm{E}[I] = \frac{\sum_{i=1}^{n-k}(\lambda_i - \overline{\lambda}) \cdot \eta_i^2}{\sum_{i=1}^{n-k} \eta_i^2} = \frac{u_*}{v} . \tag{9.7}$$

As before, the moments of $I - \mathrm{E}[I]$ are the ratio of the moments of u_* and v. The cumulants of u_* are

$$\kappa_p(u_*) = 2^{p-1}(p-1)! \cdot \sum_{i=1}^{n-k}(\lambda_i - \overline{\lambda})^p . \tag{9.8}$$

The central moments of the numerator can be obtained from the cumulants in equation (9.8) (see Stuart and Ord, 1994, Eq. 3.38). In particular, the first four moments of u_* are $\mathrm{E}[u_*] = \kappa_1(u_*) = 0$, $\mathrm{E}[u_*^2] = \kappa_2(u_*)$, $\mathrm{E}[u_*^3] = \kappa_3(u_*)$, and $\mathrm{E}[u_*^4] = \kappa_4(u_*) + 3 \cdot \kappa_2^2(u_*)$.

9.1.3   Eigenvalue or Trace Expressions for the Central Moments

With the help of the theorem by Pitman and the building blocks for the expectations of the numerator and denominator of Moran's I under the zero hypothesis of spatial independence, the central moments μ_p of Moran's I can be given in terms of the eigenvalues:

$$\left. \begin{aligned}
\mathrm{E}[I] &\equiv \frac{\sum_{i=1}^{n-k}\lambda_i}{n-k} = \overline{\lambda} , \\
\mu_2 &\equiv \frac{2 \cdot \sum_{i=1}^{n-k}(\lambda_i - \overline{\lambda})^2}{(n-k) \cdot (n-k+2)} , \\
\mu_3 &\equiv \frac{8 \cdot \sum_{i=1}^{n-k}(\lambda_i - \overline{\lambda})^3}{(n-k) \cdot (n-k+2) \cdot (n-k+4)} , \\
\mu_4 &\equiv \frac{48 \cdot \sum_{i=1}^{n-k}(\lambda_i - \overline{\lambda})^4 + 12 \cdot \left(\sum_{i=1}^{n-k}(\lambda_i - \overline{\lambda})^2\right)^2}{(n-k) \cdot (n-k+2) \cdot (n-k+4) \cdot (n-k+6)} .
\end{aligned} \right\} \tag{9.9}$$

Higher central moments of Moran's I can be generated by using the cumulant expressions in Stuart and Ord (1994, p 89, Eq. 3.38) and the equations (9.4) and (9.8). Note that the trace formulations in equations 10 and 11 of Durbin and Watson (1950) are not correct.

The expressions of Moran's I's moments in terms of the eigenvalues of $\mathbf{K} \equiv \mathbf{M} \cdot \frac{1}{2}(\mathbf{V} + \mathbf{V}^T) \cdot \mathbf{M}$ are useful if we are going to calculate the exact distribution under the zero hypothesis in the first place. If, on the other hand, we are only interested in the moments, the resource demanding evaluation of eigenvalues can be by-passed by making use of the trace operator and noting that

$$\sum_{i=1}^{n-k} \lambda_i^p = \mathrm{tr}(\mathbf{K}^p) . \tag{9.10}$$

However, for large p taking the p-th potency of $\mathbf{K}$ becomes more computing time demanding than evaluation of the spectrum of eigenvalues. Experience with numerical experiment indicates that the threshold potency for a feasible evaluation of the moments in terms of the trace is around $p = 4$

Henshaw (1966, with corrections 1967) has followed this avenue and developed the moments of ratios in quadratic forms in terms of the trace formulation. Applying equation (9.10) onto equations (9.9) Henshaw gets the central moments of Moran's I as:

$$
\left.
\begin{aligned}
\mathrm{E}[I] &\equiv \frac{\mathrm{tr}(\mathbf{K})}{n - k} \\
\mu_2 &\equiv \frac{2 \cdot \{(n - k) \cdot \mathrm{tr}(\mathbf{K}^2) - \mathrm{tr}(\mathbf{K})^2\}}{(n - k)^2 \cdot (n - k + 2)} \\
\mu_3 &\equiv \frac{8\{(n - k)^2 \mathrm{tr}(\mathbf{K}^3) - 3(n - k)\mathrm{tr}(\mathbf{K})\mathrm{tr}(\mathbf{K}^2) + 2\mathrm{tr}(\mathbf{K})^3\}}{(n - k)^3 \cdot (n - k + 2) \cdot (n - k + 4)} \\
\mu_4 &\equiv 12 \cdot \{(n - k)^3 \cdot [4 \cdot \mathrm{tr}(\mathbf{K}^4) + \mathrm{tr}(\mathbf{K}^2)^2] - \\
&\quad 2 \cdot (n - k)^2 \cdot [8 \cdot \mathrm{tr}(\mathbf{K}) \cdot \mathrm{tr}(\mathbf{K}^3) + \mathrm{tr}(\mathbf{K}^2) \cdot \mathrm{tr}(\mathbf{K})^2] + \\
&\quad (n - k) \cdot [24 \cdot \mathrm{tr}(\mathbf{K}^2) \cdot \mathrm{tr}(\mathbf{K})^2 + \mathrm{tr}(\mathbf{K})^4] - 12 \cdot \mathrm{tr}(\mathbf{K})^4\} \div \\
&\quad \{(n - k)^4 \cdot (n - k + 2) \cdot (n - k + 4) \cdot (n - k + 6)\} \ .
\end{aligned}
\right\} \quad (9.11)
$$

The variance of Moran's I is given by $\mathrm{Var}[I] \equiv \mu_2$, the skewness by $\beta_1 \equiv \frac{\mu_3}{(\mu_2)^{\frac{3}{2}}}$, and the kurtosis by $\beta_2 \equiv \frac{\mu_4}{(\mu_2)^2}$.

9.2   The Conditional Moments of Moran's I

Equation (7.6) gave the functional form of Moran's I under the presence of a hypotheticial spatial process as a ratio of quadratic forms in an independent normal distributed random vector $\boldsymbol{\delta}$. The moments of Moran's I under the assumption of spatial independence lose their legitimacy under the influence of a spatial process because Moran's I is no longer independent of its denominator (see Section 9.1.1). Direct procedures to acquire exact moments of general ratios of quadratic forms by means of their moment generating function are far less known and applied than, for instance, methods to calculate their exact distribution.

As in Section 8.2.1 the moments of Moran's I can be approximated for a given spatial process by an indirect approach based on the exact distribution. To do so, a sufficient number of equally spaced significance points within its feasible range needs to be evaluated. From these significance points the density function of Moran's I can be roughly specified which in turn allows one to calculate its moments. While this indirect procedure is numerically demanding due to the large number of evaluations of the distribution function it can be used independently from the direct procedure to cross-check its accuracy and proper implementation. This control instrument has been applied in this book to verify the direct evaluations of the moments.

Sawa (1978), based on his previous work (Sawa, 1972) on the moment generating function of a ratio of quadratic forms, was the first to give a direct approach. Let us define the matrices

$$\mathbf{A} \equiv \boldsymbol{\Omega}^{T\frac{1}{2}} \cdot \mathbf{M} \cdot \tfrac{1}{2} \cdot (\mathbf{V} + \mathbf{V}^T) \cdot \mathbf{M} \cdot \boldsymbol{\Omega}^{\frac{1}{2}} \text{ and } \mathbf{B} \equiv \boldsymbol{\Omega}^{T\frac{1}{2}} \cdot \mathbf{M} \cdot \boldsymbol{\Omega}^{\frac{1}{2}} . \quad (9.12)$$

Due to the rank defect of the projection matrix $\mathbf{M}$ the matrix $\mathbf{A}$ has at the most rank $n - p$ and the matrix $\mathbf{B}$ has rank $n - k$. Consequently, $\mathbf{B}$ is positive semi-definite. Moran's I, under the presence of a significant spatial process, can then be written as

$$I = \frac{Q_1}{Q_2} = \frac{\boldsymbol{\delta}^T \cdot \mathbf{A} \cdot \boldsymbol{\delta}}{\boldsymbol{\delta}^T \cdot \mathbf{B} \cdot \boldsymbol{\delta}} , \quad (9.13)$$

where $\boldsymbol{\delta} \sim \mathcal{N}(0, \sigma^2 \cdot \mathbf{I})$. Ergo, the joint moment generating function of Q_1 and Q_2, i.e., $\phi(t_1, t_2) = |\mathbf{I} - 2 \cdot t_1 \cdot \mathbf{A} - 2 \cdot t_2 \cdot \mathbf{B}|^{-\frac{1}{2}}$, can be re-written as

$$\phi(t_1, t_2) = |\mathbf{I} - 2 \cdot t_1 \cdot \mathbf{H} - 2 \cdot t_2 \cdot \boldsymbol{\Lambda}|^{-\frac{1}{2}} \quad (9.14)$$

with

$$\mathbf{H} \equiv \mathbf{P}^T \cdot \mathbf{A} \cdot \mathbf{P} \quad \text{and} \quad \boldsymbol{\Lambda} \equiv \mathbf{P}^T \cdot \mathbf{B} \cdot \mathbf{P} \quad (9.15)$$

where $\boldsymbol{\Lambda}$ is a $n \times n$ diagonal matrix of eigenvalues λ_i of $\mathbf{B}$ and where $\mathbf{P}$ is a $n \times n$ matrix whose columns are normalized eigenvectors of $\mathbf{B}$. Because of the rank defect, k eigenvalues in $\boldsymbol{\Lambda}$ must be zero. Therefore, the remaining $n - k$ eigenvalues and their associated eigenvectors have to be re-sequenced so that $0 < \lambda_1 \leq \lambda_2 \leq \cdots \leq \lambda_{n-k}$. Note that all eigenvalues are positive because the covariance matrix $\boldsymbol{\Omega}$ is positive definite. Following a hint by Smith (1989, p 252) for Durbin and Watson type statistics the symmetric matrix $\mathbf{H}$ decomposes into

$$\mathbf{H} = \begin{pmatrix} \mathbf{H}^*_{(n-k) \times (n-k)} & \mathbf{0}_{(n-k) \times k} \\ \mathbf{0}_{k \times (n-k)} & \mathbf{0}_{k \times k} \end{pmatrix} \quad (9.16)$$

due to the joint presence of the projection matrix $\mathbf{M}$ in the numerator and denominator. It forces both the eigenvector matrix $\mathbf{P}$ and the numerator $\mathbf{A}$ to be orthogonal to the design matrix $\mathbf{X}$. Therefore, we only need to concentrate on the $(n - k) \times (n - k)$ sub-matrix $\mathbf{H}^*$.

The conditional expectation of Moran's I can be evaluated by the improper integral

$$\mathrm{E}[I \mid \boldsymbol{\Omega}] = \int_0^\infty \theta(t, \boldsymbol{\Lambda}) \cdot \sum_{i=1}^{n-k} \frac{h^*_{ii}}{1 + 2 \cdot \lambda_i \cdot t} \cdot dt \quad (9.17)$$

where

$$\theta(t, \boldsymbol{\Lambda}) = \prod_{i=1}^{n-k} (1 + 2 \cdot \lambda_i \cdot t)^{-\frac{1}{2}} \quad (9.18)$$

with h^*_{ii} the diagonal elements of $\mathbf{H}^*$. The integrand $f(\cdot; t)$ in (9.17) is either monotonic increasing or decreasing and approaches zero rapidly, i.e.,

$\lim_{t \to \infty} f(\,\cdot\,;t) = 0$. Because the upper limit of this integral is infinite, an upper truncation point $\mathcal{T}_1$ needs to be determined for which $\int_{\mathcal{T}_1}^{\infty} f(\,\cdot\,;t) \cdot dt \leq \varepsilon$. De Gooijer (1980) approximates this upper bound by

$$\int_{\mathcal{T}_1}^{\infty} \frac{(n-k) \cdot h_{\max}^{(1)}}{(2 \cdot \lambda_1 \cdot t)^{\frac{n-k}{2}}} \cdot dt = \frac{(n-k) \cdot h_{\max}}{(2 \cdot \lambda_1)^{\frac{n-k}{2}}} \cdot \left(\frac{n-k}{2} - 1 \right) \cdot \frac{1}{\mathcal{T}_1^{\frac{n-k}{2} - 1}} < \varepsilon \quad (9.19)$$

where $h_{\max}^{(1)} = \max_{i \in \{1,\dots,n-k\}} |h_{ii}^*|$.

The variance of Moran's I under an alternative hypothesis can not be given immediately. However, Sawa (1978) gives its crude second moment as

$$\mathrm{E}[I^2 \mid \Omega] = \int_0^{\infty} \theta(t, \Lambda) \cdot \sum_{i=1}^{n-k} \sum_{j=1}^{n-k} \frac{h_{ii}^* \cdot h_{jj}^* + 2 \cdot (h_{ij}^*)^2}{(1 + 2\lambda_i t) \cdot (1 + 2\lambda_j t)} \cdot t \cdot dt \quad (9.20)$$

where $\theta(t, \Lambda)$ is given in equation (9.18). Following again De Gooijer (1980) the upper truncation point $\mathcal{T}_2$ for this infinite integral of the crude second moment can be approximated at a given precision ε by

$$\begin{aligned}
&\int_{\mathcal{T}_2}^{\infty} \frac{3 \cdot (n-k)^2 \cdot (h_{\max}^{(2)})^2}{(2 \cdot \lambda_1 \cdot t)^{\frac{n-k}{2}}} \cdot dt \\
&= \frac{3 \cdot (n-k)^2 \cdot h_{\max}^{(2)}}{(2 \cdot \lambda_1)^{\frac{n-k}{2}}} \cdot \left(\frac{n-k}{2} - 2 \right) \cdot \frac{1}{\mathcal{T}_2^{\frac{n-k}{2} - 2}} < \varepsilon
\end{aligned} \quad (9.21)$$

where $h_{\max}^{(2)} = \max_{i \in \{1,\dots,n-k\}, j \in \{1,\dots,n-k\}} |h_{ij}^*|$. In my calculations I found the resulting truncation points of $\mathcal{T}_1$ and $\mathcal{T}_2$ in equations (9.19) and (9.21) overly conservative and I suggest, therefore, using in the denominator $\frac{1}{n-k} \sum_{i=1}^{n-k} \lambda_i$ instead of λ_1. I have used Simpson's algorithm as implemented in GAUSS (Adaptec System, Inc., 1996) to evaluate the integrals (9.17) and (9.20) with my implementation of the truncation points. Calculation of the conditional expectations of Moran's I is performed by Sawa's direct method within few seconds.

De Gooijer (1980), following Sawa's derivation (1972 and 1978), also gives formulas for the exact moments and their truncation point of up to the order four. Their complexity increases with increasing order but they still remain unidimensional integrals. Further discussions and applications of Sawa's method can be found in Jones (1987) and Ali (1984), who employs a slightly different parameterization. For least-squares estimators of the autocorrelation coefficients, which can be expressed as ratio of quadratic forms, Magnus (1990) gives extended results to obtain the moments of their forecast bias and their mean-square forecast error. Smith (1989) uses a different approach. He applies zonal polynomials and invariant polynomials with multiple matrix arguments to calculate the moments of ratios of quadratic forms in normal variables. Lieberman (1994) provides a Laplace approximation to these moments of ratios of quadratic forms which can be applied also on non-normal random vectors.

9.3   Correlation between Different Spatial Structures

With either formulation of the variance of Moran's I under the assumption of spatial independence or conditional to the forces of a spatial process we can express the correlation between two Moran's I's defined on different spatial link matrices. Let I_A and I_B be two Moran's I's which differ only in the spatial link matrices $\mathbf{A}$ and $\mathbf{B}$ (i.e., they are defined on the same regression model). Both spatial link matrices are assumed to be of order $n \times n$, symmetric and non-reflective. Then the difference in both Moran's I's can be expressed as

$$I_A - I_B = \frac{\boldsymbol{\delta}^T \cdot \boldsymbol{\Omega}^{T\frac{1}{2}} \cdot \mathbf{M} \cdot (\mathbf{A} - \mathbf{B}) \cdot \mathbf{M} \cdot \boldsymbol{\Omega}^{\frac{1}{2}} \cdot \boldsymbol{\delta}}{\boldsymbol{\delta}^T \cdot \boldsymbol{\Omega}^{T\frac{1}{2}} \cdot \mathbf{M} \cdot \boldsymbol{\Omega}^{\frac{1}{2}} \cdot \boldsymbol{\delta}} \tag{9.22}$$

in terms of the i.i.d. normal vector $\boldsymbol{\delta} \sim \mathcal{N}(0, \sigma^2 \cdot \mathbf{I})$. Thus using the results in Section 9.1 if $\boldsymbol{\Omega} = \sigma^2 \cdot \mathbf{I}$ or in Section 9.2 if $\boldsymbol{\Omega} \neq \sigma^2 \cdot \mathbf{I}$ the variance of $\mathrm{Var}[I_A - I_B \mid \boldsymbol{\Omega}]$ can be calculated.

By using the relation

$$\mathrm{Var}[I_A - I_B \mid \boldsymbol{\Omega}] = \mathrm{Var}[I_A \mid \boldsymbol{\Omega}] + \mathrm{Var}[I_B \mid \boldsymbol{\Omega}] - 2 \cdot \mathrm{Cov}[I_A, I_B \mid \boldsymbol{\Omega}] \tag{9.23}$$

(see for instance Mood, Graybill, and Boes, 1974, p 179) we can solve for the unknown covariance $\mathrm{Cov}[I_A, I_B]$. Therefore, the correlation between two Moran's I's based on different spatial structures can be written as

$$\mathrm{Corr}[I_A, I_B \mid \boldsymbol{\Omega}] = \frac{\frac{1}{2} \cdot \{\mathrm{Var}[I_A \mid \boldsymbol{\Omega}] + \mathrm{Var}[I_B \mid \boldsymbol{\Omega}] - \mathrm{Var}[I_A - I_B \mid \boldsymbol{\Omega}]\}}{\sqrt{\mathrm{Var}[I_A \mid \boldsymbol{\Omega}] \cdot \mathrm{Var}[I_B \mid \boldsymbol{\Omega}]}} . \tag{9.24}$$

This derivation is in line with Oden's (1984) results except he used the sum $I_A + I_B$ of two different Moran's I's. I feel for interpretation purposes the difference between two Moran's I's is more appealing.

Under the zero hypothesis direct evaluation of the correlation in terms of the trace or the eigenvalue formulation is easy to perform by means of $\mathrm{Var}[I_A]$, $\mathrm{Var}[I_B]$, and $\mathrm{Var}[I_A - I_B]$. However, under the presence of a significant spatial process in total six integrals need to be numerically evaluated, three related to $\mathrm{E}[I_A^2 \mid \boldsymbol{\Omega}]$, $\mathrm{E}[I_B^2 \mid \boldsymbol{\Omega}]$, and $\mathrm{E}[(I_A - I_B)^2 \mid \boldsymbol{\Omega}]$ and three related to $\mathrm{E}[I_A - I_B \mid \boldsymbol{\Omega}]$, $\mathrm{E}[I_A \mid \boldsymbol{\Omega}]$ and $\mathrm{E}[I_B \mid \boldsymbol{\Omega}]$. With these expectations we can express $\mathrm{Var}[I_A - I_B \mid \boldsymbol{\Omega}] = \mathrm{E}[(I_A - I_B)^2 \mid \boldsymbol{\Omega}] - (\mathrm{E}[I_A - I_B \mid \boldsymbol{\Omega}])^2$, $\mathrm{Var}[I_A \mid \boldsymbol{\Omega}] = \mathrm{E}[I_A^2 \mid \boldsymbol{\Omega}] - \mathrm{E}[I_A \mid \boldsymbol{\Omega}]^2$ and $\mathrm{Var}[I_B \mid \boldsymbol{\Omega}]$ analogously. Therefore, calculation of the correlation under an alternative hypothesis results in substantial numerical efforts.

Note that if we want to calculate the correlation between I_A and I_B it is not necessary to impose the common constraint $\sum_{i=1}^{n} \sum_{j=1}^{n} a_{ij} = \sum_{i=1}^{n} \sum_{j=1}^{n} b_{ij} = n$ on the components of the two spatial link matrices $\mathbf{A}$ and $\mathbf{B}$. Assuming $\mathbf{A}^* \equiv a \cdot \mathbf{A}$ and $\mathbf{B}^* \equiv b \cdot \mathbf{B}$ then

$$\left. \begin{aligned} \mathrm{Corr}[I_{A^*}, I_{B^*}] &= \frac{\mathrm{Cov}[I_{A^*}, I_{B^*}]}{\sqrt{\mathrm{Var}[I_{A^*}] \cdot \mathrm{Var}[I_{B^*}]}} , \\ &= \frac{a \cdot b \cdot \mathrm{Cov}[I_A, I_B]}{\sqrt{a^2 \cdot \mathrm{Var}[I_A] \cdot b^2 \cdot \mathrm{Var}[I_B]}} , \\ &= \mathrm{Corr}[I_A, I_B] , \end{aligned} \right\} \tag{9.25}$$

because the correlation is a scale-free function. Consequently, the correlation between two Moran's I's is invariant of the scales of the spatial link matrices. This result is especially useful when dealing with the correlation between local Moran's I_i's.

9.4   Some Applications Using Moran's I's Theoretical Moments

This section will briefly sketch two applications of the theoretical moments of Moran's I. The first deals with the approximation of exact distribution of Moran's I by some well-known simple distributions. This allows fast assessment of the significance of an observed Moran's I_0 without numerical evaluation of its exact probability. The second application deals with linking an observed Moran's I_0 to its related autocorrelation level ρ of either an autoregressive or a moving average spatial process.

9.4.1   Approximations of Moran's I's Distribution by its Moments

As noted in Section 8.2.2 the skewness of the normal distribution is 0 and its kurtosis is 3 so that any deviation of Moran's I's skewness and kurtosis from these values indicates a lack in normality. In theory also the moment of higher order must be considered (see the Hoeffding Theorem 10), however, for practical applications they are consistently ignored in the literature. There are no significance tests to check the deviation of the distribution of Moran's I from the normal distribution because both distributions and their moments are not random functions. Therefore, any comparison between both distributions must be conducted in analytical terms.

If the skewness and the kurtosis of Moran's I do not differ substantially from their counterparts of the normal distribution we can use the z-transformation of Moran's I

$$\frac{I_0 - \mathrm{E}[I]}{\sqrt{\mathrm{Var}[I]}} \simeq \mathcal{N}(0, 1) \tag{9.26}$$

to obtain the significance of an observed Moran's I_0. Because of the simplicity of this approximation, its use is by far the standard procedure in spatial statistics and spatial econometrics (see Cliff and Ord, 1981, Anselin, 1988, Griffith 1987 and 1988, Goodchild, 1986, Odland, 1988 and others). Theoretically it has a close resemblances to the Lagrange Multiplier test (see Chapter 4.3.1).

However, if there is a marked difference between the skewness and the kurtosis of Moran's I with that of the normal distribution alternative approximation strategies need to be employed. A direct consequence of this mismatch of both distributions is to apply the direct evaluation of the probability of an observed Moran's I_0 by Imhof's method. This is the preferred strategy under the present advances in computer algorithms and technology. Yet traditionally also approximations to the four-parameter Pearson distribution family have been extensively applied:

1. Based on Durbin and Watson's (1951) approximation Cliff and Ord (1972) model the distribution of Moran's I by the Beta-distribution. They used Moran's I's feasible range and its expectation and variance. While this approximation acknowledges that instead of the normal distribution the distribution of Moran's I is bound by its feasible range it neglects its shape expressed by the skewness and kurtosis.

2. Costanzo et al. (1983) approximated the distribution of Moran's I by a Pearson Type III distribution. The major drawback of this approximation is that it assumes a fixed linear relationship between the skewness and the kurtosis.

3. Hepple (1998) followed the procedure outline by Henshaw (1966 and 1967) which utilizes all four moments. First the kurtosis and the skewness are used to approximate the two parameters characterizing the Beta-distribution. Using these parameters and the expectation and variance in a second step the feasible range of Moran's I is approximated. However Hepple experienced that for sparse spatial link matrices the Beta-distribution is not always a feasible model. This subject will be discussed below.

4. Ali (1984a and 1984b) used Edgeworth-type asymptotic expansions and the Pearson distribution family to compare their approximation accuracy for the distribution of the Durbin and Watson's serial autocorrelation statistic against its exact distribution. He focused on the assumption of serial independence as well as the presence of significant stationary autoregressive-moving-average processes. For the approximation of a significant serial process he used the conditional moments of the Durbin and Watson test. He concluded that only the Pearson approximation seems to be reliable for the Durbin and Watson test in relatively small samples.

Stuart and Ord (1994, Chap. 6) give a concise discussion of the Pearson distribution family of which the Beta-distribution and the normal distribution are members. In particular, under non-normal distributed situations the Beta-distribution is often used for approximation purposes because by variation of only two parameters it can take highly skewed, as well as U- and J-shaped distributional forms (see Moods el al. 1985, p 115 for examples of the Beta-distribution). However, there are algebraic limits to the feasible combinations for the skewness and the kurtosis parameters. Stuart and Ord (1994, p 216) display in Figure 6.1 these limits. For instance, if the skewness is around zero the kurtosis must be smaller than 3 to approximate the underlying distribution by a Beta-distribution. In general, sparse spatial link matrices, like higher order neighbor matrices or local link matrices, often end up with a parameter combination for which the whole set of four-parameter Pearson distributions does not offer a feasible family member. Stuart and Ord (1994, Chap. 6) also discuss other approximations, for instance, by Chebyshev-Hermite polynomials and Edgeworth's series. Terui and Kikuchi (1994) have used a Edgeworth-type asymptotic expansion to approximate the distribution of Moran's I.

9.4.2   Potential Applications of Moran's I's Conditional Expectation

Exact Methods versus Simulation Studies The conditional expectation
can be used to link an observed value of Moran's I to either the autocorrelation
coefficient ρ of a moving average or an autoregressive spatial process. So far
simulation studies have been used to establish such a link (see Anselin and Rey,
1991, Anselin and Florax, 1995, and Florax and Rey, 1995). However, one has
to be careful with generalizations of Monte Carlo studies (see also Sawa, 1978).
In particular, in the empirical spatial studies their results always depend on the
unique irregular tessellation of the spatial objects. For a detailed discussion on
this problem with simulations studies see the exchange between Haining (1986)
and Anselin (1986a and 1986b). It is much better to give a link in analytical
terms which can be directly evaluated under any changes in the underlying model
structure (Lieberman, 1994) instead of the exploitation of laborious Monte Carlo
experiments.

In time-series analysis Sawa (1972) investigated the k-class estimator in simultaneous equations models by means of the conditional moments. Furthermore, Sawa (1978) used the conditional moments to evaluate the bias of least
squares estimators of the autocorrelation coefficient and the real process autocorrelation parameter. Furthermore, he assessed the quality of approximations.
De Gooijer (1980) extended the investigation by Sawa for ARIMA-processes.
Ali (1984) used the moments of sample autocorrelation coefficients under a significant ARMA-process to approximate their exact significance point by a distribution from the four-parameter Pearson family. The relation between exact
moments and Stein-type predictors in regression analysis have been discussed by
Jones (1987). Finally, Magnus (1990) used the exact moments of a least squares
estimator for the autocorrelation coefficient not only to evaluate the bias in an
AR(1)-process but also the forecast bias and the mean-square forecast error.

Moran's I's link to Gaussian Spatial Processes In spatial analysis the
conditional expectation can be used to evaluate the bias of the least squares
estimator for the autocorrelation coefficient ρ of an autoregressive model. Based
on an autoregressive error structure in the regression residuals (i.e., an autore-
gressive error model given as $\widehat{\varepsilon} = \rho \cdot \mathbf{V} \cdot \widehat{\varepsilon} + \delta$) Anselin (1980, see also Haining
1994) uses the least squares estimator for ρ as

$$\widehat{\rho}_{ols} = (\widehat{\varepsilon}^T \cdot \mathbf{V}^T \cdot \mathbf{V} \cdot \widehat{\varepsilon})^{-1} \cdot \widehat{\varepsilon}^T \cdot \mathbf{V}^T \cdot \widehat{\varepsilon} . \tag{9.27}$$

The bias of the observed estimator $\widehat{\rho}_0$ can be evaluated in line with Sawa's (1978)
approach by its conditional expectation $\widehat{\rho}_0 = \mathrm{E}\big[\widehat{\rho}_{ols} \mid \sigma^2 \cdot \big((\mathbf{I}-\rho\mathbf{V})^T \cdot (\mathbf{I}-\rho\mathbf{V})\big)^{-1}\big]$.
Using for instance a bi-section search procedure the conditional expectation can
be solved for ρ which matches the observed ordinary least squares estimator $\widehat{\rho}_0$.

Note that the numerator of Moran's I closely resembles the numerator $\widehat{\varepsilon}^T \cdot$
$\mathbf{V}^T \cdot \widehat{\varepsilon}$ of the least squares estimator $\widehat{\rho}_{ols}$ in equation (9.27). However, the denominator is defined in terms of the spatially lagged regression residuals $\mathbf{V} \cdot \widehat{\varepsilon}$,
therefore, $\widehat{\rho}_{ols}$ is unbound (Anselin, 1980, p·51). In contrast to the least squares

estimator $\widehat{\rho}_{ols}$, Moran's I is not only related to an autoregressive spatial process but also to a moving average process. King (1981, see also Chap. 4.3.2) has shown that under the assumption of spatial independence, Moran's I is a locally best invariant test against both alternative process formulations. To link Moran's I

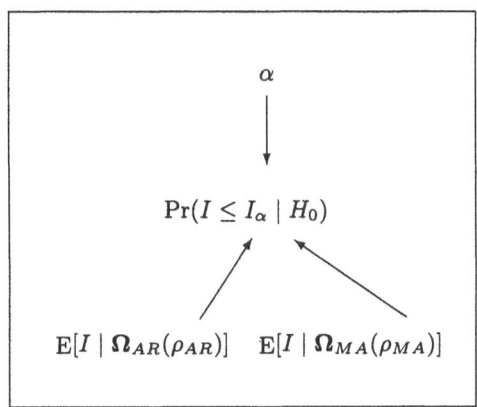

Fig. 9.1. Link between a given significance level α, its associated Moran's I_α and the matching autocorrelation levels ρ_{AR} and ρ_{MA} of Gaussian spatial processes

to either the related autocorrelation level ρ of a moving average process or the autocorrelation level ρ of an autoregressive process one must choose a critical value I_α of Moran's I which has probability $\alpha = \Pr(I \le I_\alpha \mid H_0)$ under the zero hypothesis. This critical value I_α is invariant under both process specification and possess the same error probability α. Therefore, we are allowed to tie the critical value I_α together with both conditional expectations

$$I_\alpha = \mathrm{E}[I \mid \Omega_{MA}(\rho_{MA})] = \mathrm{E}[I \mid \Omega_{AR}(\rho_{AR})] \tag{9.28}$$

of either a moving average process with an autocorrelation level ρ_{MA} or an autoregressive process with an autocorrelation level ρ_{AR}. Again these autocorrelation levels ρ_{MA} and ρ_{AR} have to be found by a bi-section search or any other search and interpolation technique like Lagranian interpolation or the secant method (see Press et al., 1989). The logic of this linking procedure is displayed in Figure 9.1. Note that in this case Moran's I_α is not a random variable, therefore, the matching procedure is theoretically guided. With the same reasoning also an observed Moran's I_0 can be linked to either the autocorrelation level ρ_{MA} or ρ_{AR}. However, since an observed value I_0 of Moran's I is a random variable, here the matching procedure is subject to random variations.

A Numerical Example In this example the hexagonal spatial tessellation of the 61 cells arranged in rings around a central cell (see Figures 7.1 or 8.4) and

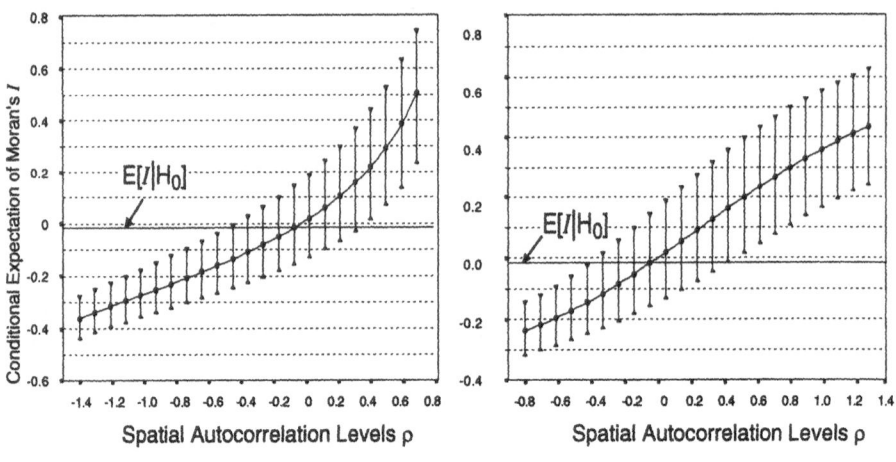

Fig. 9.2. Conditional expectations of Moran's I for different autocorrelation levels of Gaussian spatial processes

the projections matrix $\mathbf{M}_{(1)}$ are employed again. The exact feasible range of Moran's I in the C-coding scheme is $I \in\,]{-}0.5458, 0.9701[$, that of an autoregressive process is $\rho_{AR} \in\,]{-}1.8322, 0.9164[$ and that of a moving average process is $\rho_{MA} \in\,]{-}0.9164, 1.8322[$. Due to the reverse relationship between the feasible ranges of the autocorrelation parameters ρ_{MA} and ρ_{AR}, it can be expected that both spatial process and their related conditional expectation of Moran's I will show an inverse relationship, too. Figure 9.2 displays the conditional expectations of Moran's I and the 95% confidence limits derived from the conditional distributions around them. It can be seen that Moran's I reacts more sensitively to an autoregressive than to a moving average process especially for negative autocorrelation.

To link an observed Moran's I_0 to either ρ_{MA} or ρ_{AR} we have to follow horizontal from the ordinate I_0 to the graph of the conditional expectations and then find the related ρ on the abscissa. Table 9.1 gives some numerical values ρ_{MA} and ρ_{AR} related to several significance points of Moran's I_α. Due to the

Table 9.1. Critical values of Moran's I and related process autocorrelation levels ρ

$\Pr(I \le I_\alpha \mid H_0)$	Critical value I_α	AR-process with ρ_{AR}	MA-process with ρ_{MA}
0.01	-0.1777	-0.5676	-0.5277
0.05	-0.1357	-0.3989	-0.3712
$1-0.05$	0.1146	0.3157	0.3624
$1-0.01$	0.1754	0.4279	0.5302

differences in the feasible ranges of the autocorrelation levels ρ_{MA} and ρ_{AR} we observe for a given I_α smaller ρ_{AR}-value for negative autocorrelation and larger ρ_{MA}-values for positive autocorrelation.

Part III

An Empirical Application of Moran's *I*'s Exact Conditional Distribution

The purpose of this Part is to demonstrate the feasibility and flexibility of the proposed statistical methodology for spatial process identifications and explorations (i) in an irregular spatial tessellation representative in size for empirical problems and (ii) to apply it to a case study taken from spatial epidemiology. Common to all examples is a regression model based on a data-set[1] of age-standardized male urinary bladder cancer incidence rates and a set of confounding and potentially causal variables. These data are available for the 219 counties of the former German Democratic Republic (GDR) aggregated over a 15 year period. The regression model is specified to capture the global variation in the incidence rates. Its residuals are initially screened for model violations such as heteroscedasticity, non-normality and global spatial autocorrelation. Furthermore, different spatial link matrices are developed to model and identify an underlying hypothetical spatial process in the disturbances. A simple spatial link matrix leads to the classical misspecification assumption that missing exogenous factors or spatial aggregation tie adjacent unexplained variations together. The underlying adjacency structure rests on a definition in absolute space. The other link matrix is specified on theoretical grounds and relates the observed disease rates in part to an underlying interregional migration graph which is embedded in a relative space. Due to the long latency period of cancers we are allowed to match the disease rates with a migration process. If in fact migration effects are present, the observed local disease rates as well as the local behavioral factors are blurred by immigration and out-migration. Consequently, both are only remotely related to the underlying location specific risk factors.

In the *first application* the inherent spatial autocorrelation based on a neighborhood relationship is analyzed with the proposed methodology for a spatial link matrix in the variance stabilizing S-coding scheme. The underlying autocorrelation level of an autoregressive spatial process is identified by means of the conditional expectation. This allows me to investigate the local heterogeneity and non-stationarities in the global spatial process and to tag local clusters of outstanding positive autocorrelation in either negative or positive regression residuals as well as significant local outliers with respect to their neighborhood reported by negative local autocorrelation. The conditional distribution of the local Moran's I_i test statistic used for this analysis and its properties are discussed. Also its conditional distribution is compared against its distribution under the assumption of spatial independence. Negligence of a significant spatial process leads to a biased interpretation of the spatial clusters and outliers.

The *second application* aims at the identification of the underlying spatial process which might have been generated in the regression disturbances. This process is supposed to be based either on the adjacency or a migration structure. This identification is done by the power function and the conditional significance of global Moran's I subject to an underlying reference process. An unambiguous

---

[1] The incidence rates following a world age-standardized form are tabulated for districts of the former GDR in Mehnert et al. (1992). Furthermore, age-standardized death rates for several cancers in each districts of the Federal Republic of Germany can be found in atlas of cancer mortality (Becker and Wahrendorf, 1997).

decision rule is designed by mutually comparing the spatial adjacency processes against the spatial migration process. This can be done because in contrast to the time-series situation substantial flexibility is gained in spatial analysis: the structure of an underlying spatial reference process can differ from the spatial structure under question on which Moran's I is defined. This way, hypotheses expressed in the form of a specific spatial structure can be evaluated by means of Moran's I's conditional distribution under the presence of a significant spatial process attached to an alternative spatial structure.

Chapter 10

The Basic Ecological Model and Spatial Setting

The first chapter of this empirical part lays the foundation of the subsequent spatial analysis by describing the characteristics of the epidemiological data set used and the spatial setting of the former GDR. In contrast to the standard disease mapping procedures, which map standardized incidence rates or smoothed functions of them directly (see Marshall, 1991, for a review on Bayesian smoothing techniques in spatial epidemiology), I will concentrate on the spatial distribution of regression residuals of an epidemiological model. Recall that systematic behavior in these residuals with respect to the underlying spatial tessellation can be interpreted in two different ways: (i) classical disease mapping searches for *missing risk factors* operating on a larger spatial scale which tie the observations together into clusters, or (ii) an underlying *spatial interaction process* which imposes some degree of organization on the observations.

Regression residuals from an ordinary least squares regression model are used in the following examples. The focus lies on a model of the spatial variability in male urinary bladder cancer incidence rates for the 219 counties of the former GDR. *Incidence rates*, unlike *mortality rates*, come with the advantage that they account also for the survivors who have been afflicted by a potentially fatal disease. This is important since the relative 5-year male survival rate for bladder cancer in the former GDR was 28.6% in 1978–1979 with a rising tendency. Note that in some Western societies the relative 5-year survival rate for bladder cancer has reached more than 70% already (see Schön et al., 1995, p 290) which, however, is primarily due to enhanced early-diagnostic methods. In part these diagnostic capabilities also explain why the observed bladder cancer incidence rates in Western Europe are higher than those in Eastern Europe (see, for instance, Schön et al., 1989, p 355 for a comparison with world standardized incidence rates). Schön and Bellach (1995) note that the "National Cancer Registry" of the former GDR is one of the most systematic and complete data sets on cases of cancer in the world. For various political and technical reasons, this data source has not been exploited for epidemiological studies in the former GDR.

In a first exploratory analysis, this data set has been probed in an atlas of cancer incidence by Mehnert et al. (1992). For male bladder cancer, as for most other cancers probed, these authors detected a marked difference between incidence rates in urban and rural counties, with an increased risk for urban counties. This *urban/rural dichotomy* in cancer incidence rates has been observed in several studies (Glick, 1982, Greenberg, 1983, Nasca et al., 1992, and Walter, 1993) and points to environmental, occupational, socio-economic and life-style differences. In part, also the diffusion of innovations in the histological assessment of tissue samples from medical schools at university cities to county hospitals may contribute to this urban/rural dichotomy. As Schön el al. (1995, p 287) note, in the eighties histopathologists started re-classifying bladder papilloma as malignant. Other risk factors with etiologic significance are smoking, late effects of exposure to radiation, work in rubber, leather, dyestuffs, textile, iron, and metal industries and as a hairdresser. Furthermore, Mehnert et al. (1992) observe a tendency for most age-standardized cancer incidences to cluster spatially.

10.1   The Setting of the Regression Model

In detail the endogenous variable male urinary bladder cancer (ICD9 188[1]) has the following characteristics: (i) the incidence rates have been aggregated into a 15-year (1974–1988) annual average; (ii) on average 19.8 new cases per $100,000$ inhabitants per year with a moderate standard deviation of 3.81 were observed; (iii) virtually no cases are observed in the male population younger than 35 years; however, the age-specific rate increases sharply for males over 45 and reaches its peak in the age group 75–85 with 125 new cases per $100,000$ persons and year (see Figure 10.1); and (iv) to adjust for the county specific age-distributions the incidence rates have been directly standardized (see Meade et al., 1988, pp 55–58) to the European reference age distribution (Doll et al., 1970 and Figure 10.2). The selection of the European age distribution is vital for the analysis of the disease rates in industrialized societies such as the former GDR and post-industrialized societies. It concentrates on the critical older population segments (see Figure 10.1) whereas the World reference population distribution emphasizes the younger population segment. Stochastic fluctuations in the incidence rates are alleviated widely because bladder cancer is not a rare disease and the data are highly aggregated on the temporal and spatial scale.

To screen the regression residuals later on several confounding and potential explanatory variables were considered. Negligence of confounding variables can override any other aspect of the methodology (Wartenberg and Greenberg, 1993). Besides the constant term, the following four exogenous variables are included in the model: (i) the z-transformed longitude of the county centroids to capture a west-east gradient[2] in the incidences as can be seen from Figure 10.3; (ii) the well-known risk-factor smoking;[3] (iii) as a proxy-variable the natural logarithm

---

[1] In the 9-th revision of the international disease classification the codes 140–208 denote cancers. In particular, the code 188 indicates urinary bladder cancer.

[2] The use of the longitude makes sense under the perspective that the former GDR was framed in-between West and East Europe. Both West and East Europe are known to show marked differences in their disease patterns for many cancers. An initial consideration of the latitude was justified by the geographic differences of the former GDR ranging from glacial lowlands in the North over loess lowlands and hills to medium and high altitude areas in the South. However, the latitude turned out to be not significant. – From a statistical point the data need to be tested for non-stationarities with respect to a trend surface.

[3] The age-standardized county specific male-smoker rate has been used here as it was defined by Schön and Bellach (1995).

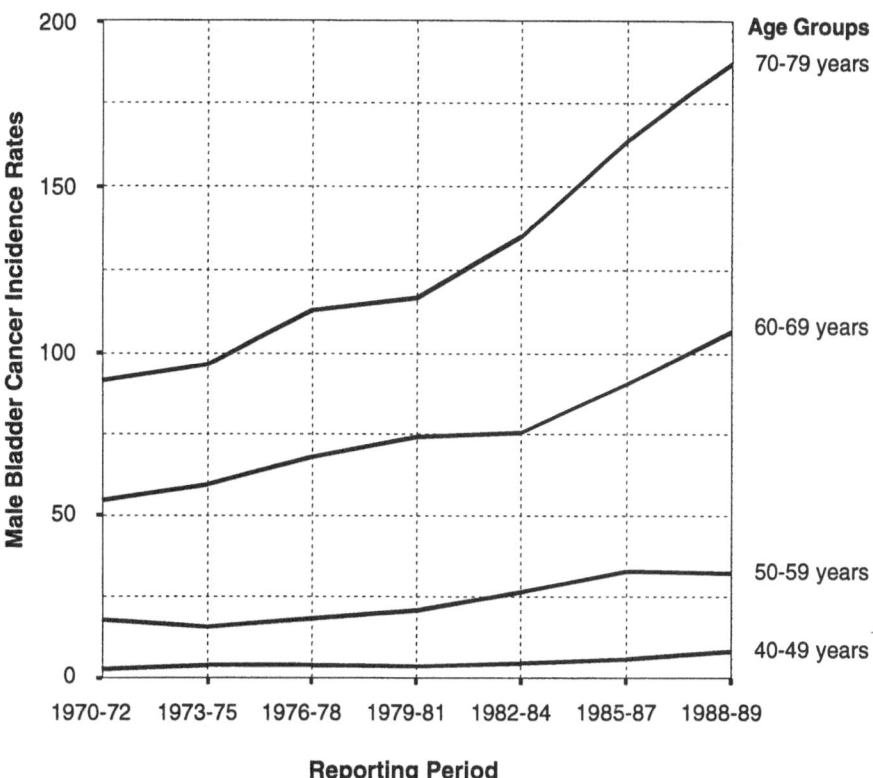

Fig. 10.1. Increasing age-specific incidence rates for male bladder cancer in the former GDR. Source: Robert Koch Institute in Berlin, Germany (http://www.rki.de/chron/dachdok/ergebniss/harnbl.htm)

of the population density to model the urban/rural dichotomy; and (iv) for technical reasons, an urban/rural dummy variable. Figure 10.3 denotes the urban counties which mostly show outstanding incidence rates. The urban/rural dummy variable is used so that the expectations of the residuals within the urban and rural spatial subsystems are both zero. This adjustment is necessary to capture potential non-stationarities with respect to the different mean levels in the spatial hierarchy. A hierarchical spatial interaction process is employed later in this analysis. In essence, Haining (1981) introduced this approach to model these non-stationarities in a central place system. He found that careful inclusion of the different levels within a central place system led to a well-specified model which shows no significant spatial autocorrelation with respect to the spatial hierarchy thus eliminating a misspecification bias.

Table 10.1 displays the estimated parameters and t-values of the regression model. With an $R^2 = 0.165$, the overall fit of the model is poor. The double count of the effects of the urban/rural dichotomy by a dummy variable and

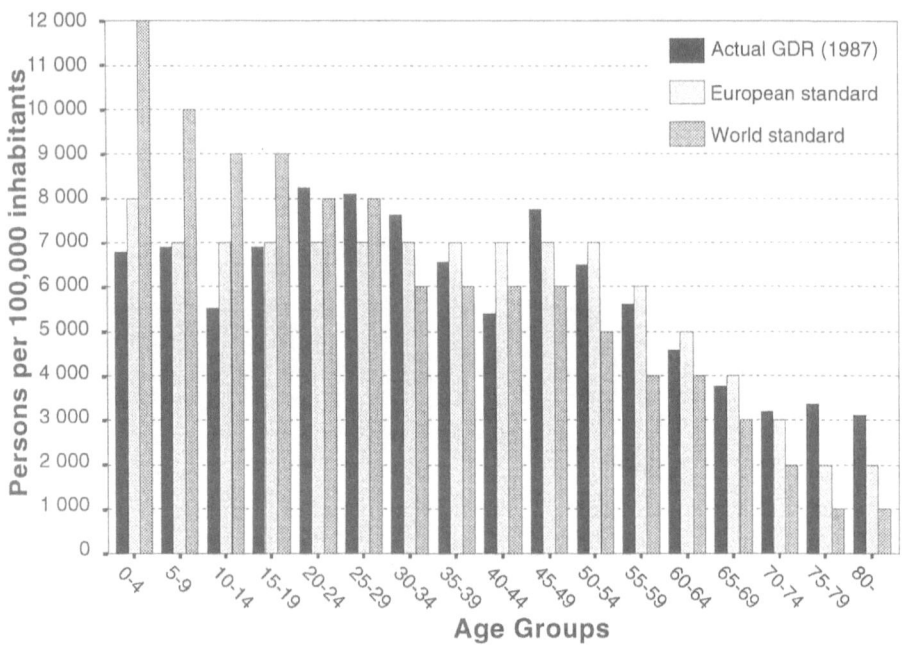

Fig. 10.2. Comparison of the age distribution in the former GDR (1987) with the European and world standard age distributions

by the population density may cause multicollinearity. The tolerance[4] values in Table 10.1 indicate a slight presence of multicollinearity. However, it does not affect the model in any critical manner. The overall effect of this multicollinearity is that the t-values of the regression parameters are decreased slightly, as can be seen by comparing this model with the one given in Tiefelsdorf and Boots (1997). Despite the fact that the urban/rural dummy is not significant it should be kept in the model to force the residuals to be unbiased with respect to their level in the spatial hierarchy. All signs of the regression parameters point in the postulated direction and the behavioral risk factor 'smoking' has the strongest impact on the propensity to develop bladder cancer.

Due to the high spatial and temporal integration of the data, an analysis of the regression residuals shows no significant departure from the normal distribution. The Kolmogorov-Smirnov test has a two-sided tail probability of $\alpha = 0.38$. Therefore, it does not deviate substantially from the zero hypothesis. Moreover, a likelihood ratio test on multiplicative heteroscedasticity (see Fomby et al.,

---

[4] The tolerance is a statistic used in regression analysis to determine multicollinearity. It measures in how far the independent variables are linearly related to one another. It is calculated as 1 minus the R^2 for an independent variable when it is predicted by the other independent variables already included in the analysis. SPSS regards tolerance values of single variables below 0.0001 as critical.

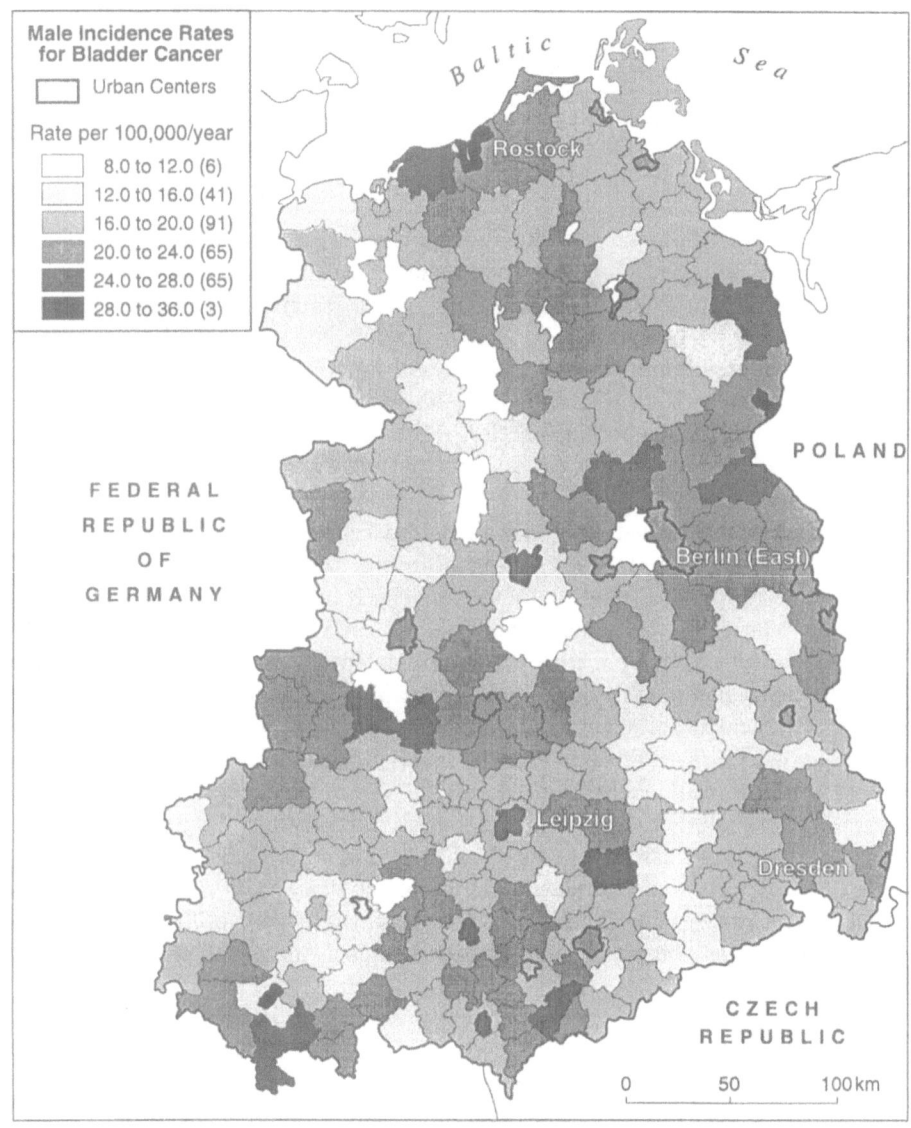

Fig. 10.3. Urban counties and age-standardized incidence rates of male bladder cancer in the former GDR

Table 10.1. Specification and estimated parameters of the underlying regression model for male urinary bladder cancer

exogenous variable	regression coefficient	tolerance	t-value
z(longitude)	0.482	0.919	1.949
male smoke-rate	0.221	0.802	2.973
ln(Pop.-density)	0.928	0.280	2.292
urban/rural dummy	1.377	0.294	1.055
constant	0.386	–	0.069

1984, p 177) based on the local incidence rates divided by the local population at risk (males 30 and older) does not indicate a significant disagreement with the constant variance assumption. Assuming the variance of the spatial objects follows $\sigma_i^2 = \sigma^2 \cdot (\text{IncidenceRate}_i/\text{PopAtRisk}_i)^\phi$, for a weighted least squares regression the maximum of the log-likelihood function is -581.381 at $\phi = 0.31$. Under homoscedasticity (i.e., $\phi = 0.0$) its value is -582.682. Thus twice the difference of the log-likelihood function values (i.e., 2.602) is not significant. At $\alpha = 0.05$ the critical value of the χ^2-distribution with one degree of freedom is 3.84.

Without being formal, this lack of heteroscedasticity is not surprising. As stressed by Pocock et al. (1981), the variance of the spatially distributed incidence rates is supposed to be composed additively of a location specific *sampling variation* related to the underlying binomial distribution and a global *unexplained variation* associated with the regression to model them. Due to the poor fit of the regression model, the global unexplained variation dominates the summation. Consequently, this example satisfies two crucial assumptions of ordinary least squares. The assumption of spatial independence in the regression residuals will be tested in the next section. However, even under the presence of positive spatial autocorrelation the observed heteroscedasticity test result is on the safe side. As Anselin and Griffith (1988) have shown in a simulation experiment, under positive spatial autocorrelation, heteroscedasticity tests have the tendency to over-reject the zero hypothesis. Note however, that their simulation results are inherently heteroscedastic because the S-coding scheme of the spatial structure has not been used.

10.2   The Spatial Setting

First, this section will give some reasons why we can expect even for noninfectious diseases an underlying spatial process. Then it will describe the population distribution in the former GDR and move on to the population dynamics which allow the reconstruction in a first approximation of the main migration pattern between the counties. This pattern resembles the hierarchical settlement system of the former GDR and can be encoded in relative space in a hierarchical graph.

10.2.1   Rationale for an Underlying Spatial Process in Disease Rates

Migration of the population under risk and a long *latency period* of a disease are often supposed to blur the association between potential risk factors and the place of residence where the outbreak of a disease is finally registered (King, 1979, and Marshall, 1991). For many cancers with a latency period of up to twenty years and more this temporal lag between exposure and outbreak increases the odds for a spatial lag as well. Knowledge of the structure of these temporal and spatial lag relations allows some of this uncertainty to be captured. As pointed out by Anselin (1988, p 42), the specific form of a spatial process should be determined a priori, presumably based on substantive theoretical grounds. Figure 10.4 sketches for two regions the theoretical basis of a mixing process

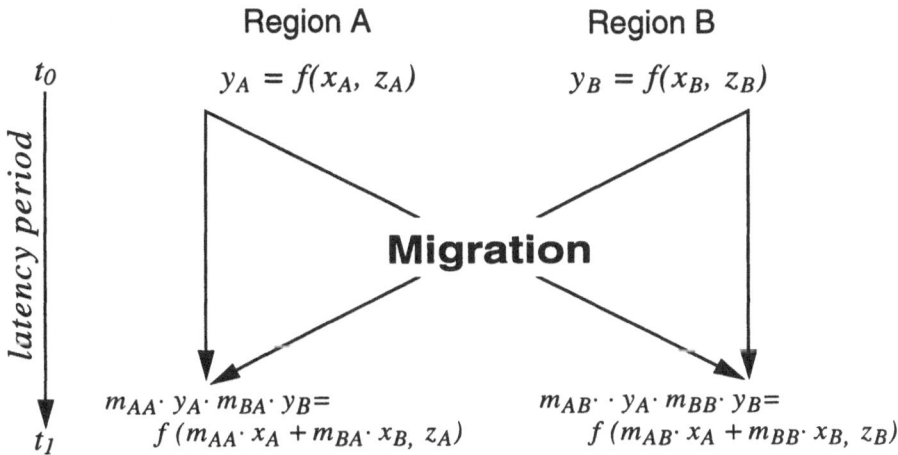

Fig. 10.4. Spatial and temporal lag relations in an ecological cancer model

due to interregional migration. Because we are dealing with an ecological cancer model, which is built from spatial aggregates and not defined on individual cases and controls, the observed and dormant local incidence rates $\{y\}$ as well as the local socio-economic and behavioral risk factors $\{x\}$ attached to the potential migrants are blurred at the registration time t_1 by the mixing migration process $\{m\}$. The longer the latency period the stronger will be the impact of this mixing process. Note that as indicated in Chapter 4.1.1 some temporal arguments have to be employed to move from the plain misspecification perspective toward a full-fledged spatial process.

The joint impact of the mixing process on the incidences $\{y\}$ and the risk factors $\{x\}$ makes it reasonable to assume that both are autoregressively linked. However, note that the environmental risk factors $\{z\}$ are region specific, so that they are not affected by this mixing process $\{m\}$. This leads to an interwoven specification of a spatial autoregressive and spatial lag model (see Chapter 4.2.3) in form of

$$\mathbf{y} = \rho \cdot \mathbf{V} \cdot \mathbf{y} + (\mathbf{I} - \rho \cdot \mathbf{V}) \cdot \mathbf{X} \cdot \boldsymbol{\beta} + \mathbf{Z} \cdot \boldsymbol{\gamma} + \boldsymbol{\eta} \,. \tag{10.1}$$

Nevertheless, because the behavioral component is usually more important to develop a degenerative disease, as a first approximation an overall autoregressive specification of the spatial process defined on the dominant migration pathways $\mathbf{V}$ has been selected in this monograph by explicitly assuming that the environmental risk factors $\{z\}$ can be neglected, that is, $\mathbf{Z} \cdot \boldsymbol{\gamma} = \mathbf{0}$, or that they can be approximated by an autoregressive component $(\mathbf{I} - \rho \cdot \mathbf{V}) \cdot \mathbf{Z} \cdot \boldsymbol{\gamma}$. This assumption, which proposes an underlying autoregressive Gaussian spatial process even for noninfectious diseases, will be tested in Chapter 11.3.

This indirect process perspective is somewhat at odds with Haining's (1991, p 225) assessment arising from a small-scale urban data-set. He states that non-infectious diseases at one location are unlikely to affect neighboring regions. His perspective leads to the classical approach which specifies spatial links on the adjacency of the spatial objects (i.e., pairs of objects sharing a common boundary). As noted in Chapters 2.3.2 and 4.1.3 spatial autocorrelation in the regression residuals based on an adjacency graph may result, for instance, from an inadequately specified regression model or missing risk factors and neglected confounding variables operating on a scale which overlaps the spatial enumeration units. Other potential short range linking factors are, for instance, a genetic transmission mechanism due to inter-district marriages or spill-over effects induced by work place commuting across the boundaries of the statistical units.

10.2.2   Approximation of a Hierarchical Migratory Graph

Of the 219 counties of the former GDR 28 are classified as urbanized.[5] Almost one third of the 16.6 million inhabitants (1987) of the former GDR live in urban areas. On average each county has 75,000 residents with the two extremes of only 17,900 inhabitants in Röbel/Müritz (located half way between Berlin and Rostock) and Berlin (East) with a population of 1,246,900. Figure 10.5 displays the population distribution in the former GDR. Note that only 26 urban counties are highlighted by a solid pie chart because two urban areas[6] lack an independent hinterland. Therefore, Halle-Neustadt has been linked to Halle (situated directly

---

[5] Counties with a population density greater than 500 inhabitants/km^2 were classified as urbanized.

[6] These two cities were planned and built after 1950 by the communist administration in an effort to increase industrial production.

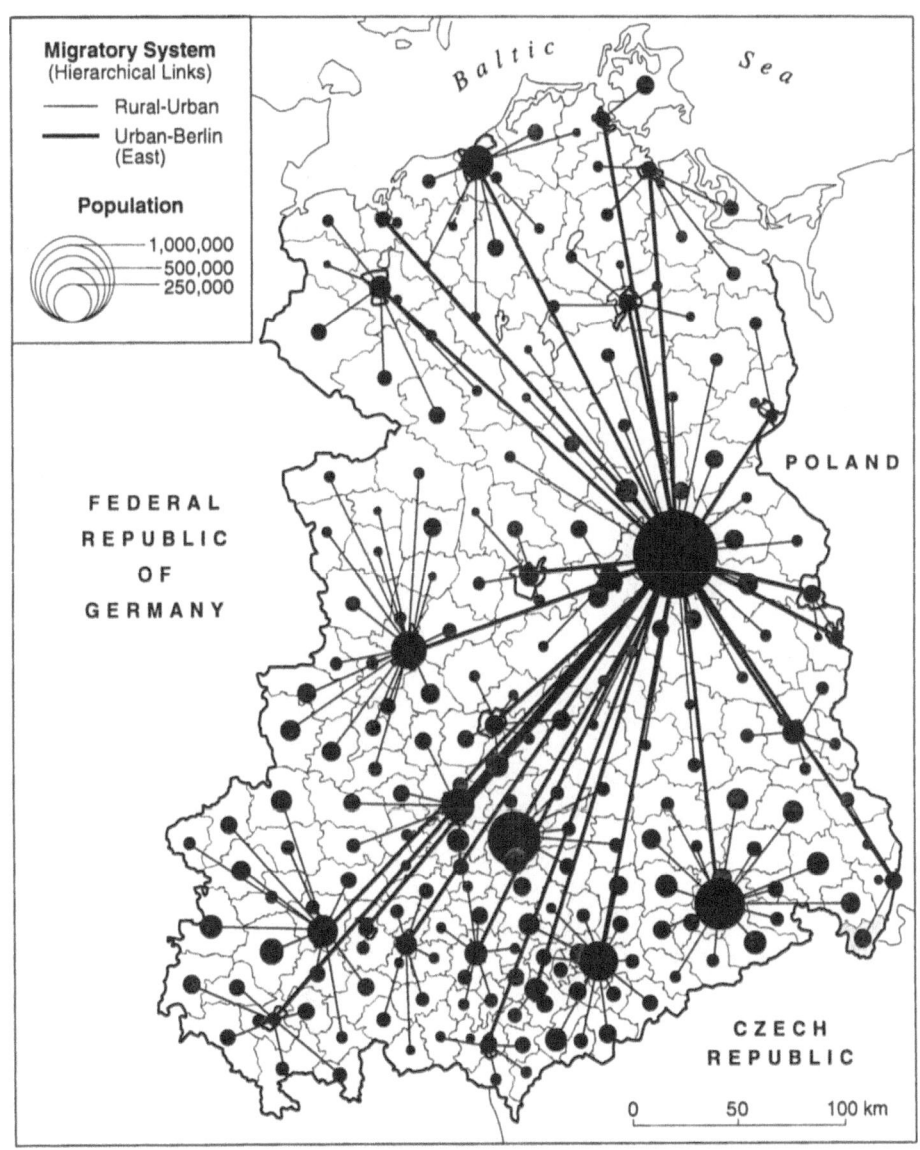

Fig. 10.5. The population base (1987) and spatial hierarchy in the former GDR

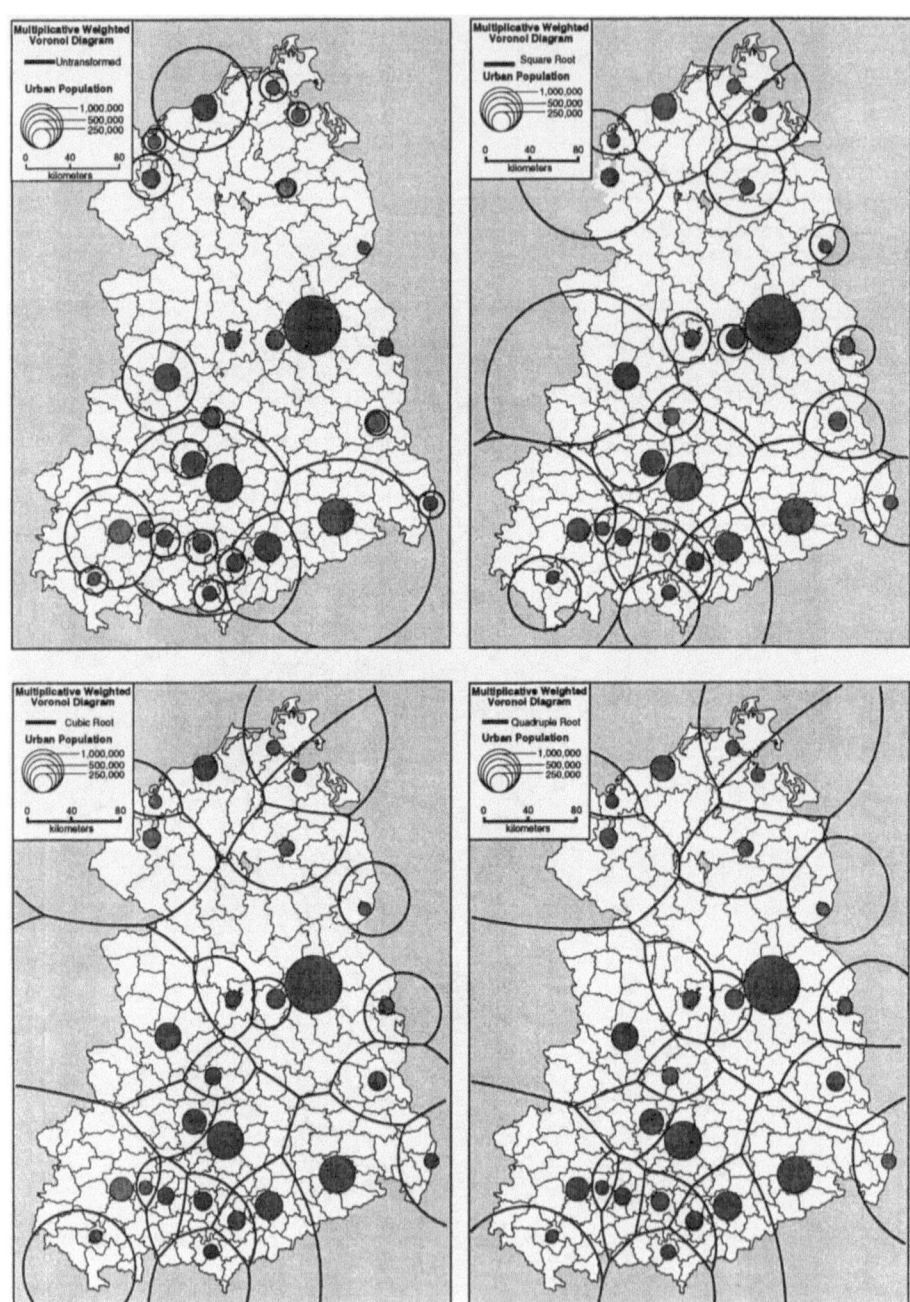

Fig. 10.6. Spheres of influence around the urban centres determined by multiplicative weighted Voronoi diagrams depending on the transformed population basis

west of Leipzig) and by the same argument Eisenhüttenstadt has been attached to Frankfurt/Oder (located directly east of Berlin).

In contrast to Western societies, migration in the former GDR was regulated indirectly by the communist administration. Free housing and labor markets were not present and the spatial allocation of the work force was centrally steered. The administration favored urban centers, and Berlin (East), the capital of the former GDR, in particular. Therefore, suburbanization and counter-urbanization virtually did not happen. Overall, the intercounty migration volume with only 2.4% per year was significantly lower than in Western societies. In general, rural counties were constantly loosing population whereas the urban counties were growing with the exception of Leipzig and a few smaller cities. The population base of these cities decreased between 1970 and 1987. Details on the migration process in the former GDR can be found in Wendt (1994), the Raumordnungs-bericht (Bundesministerium für Raumordnung, Bauwesen und Städtebau, 1993) and the statistical yearbook of the former GDR (Staatliche Zentralverwaltung für Statistik, 1988). However, to the author's knowledge, spatial interaction matrices at the county level[7] are not available. Thus, the potential pathways of the migration process have to be approximated by using aggregated aspatial characteristics.

Two assumptions (see for instance Cadwallader, 1992) underlie the *approximation of the migration pathways*: on the first level migrants follow the maximum gravitational field of the urban areas; and on the second level there are interurban links. On the first level potential migrants residing in rural counties move to the urban area with the highest gravitational field in their reach, and on the second level all urban counties have ties to Berlin (East). The gravitational potential is determined by the population base of the urban areas adjusted by their population growth rate. A multiplicatively weighted Voronoi diagram (see Okabe et al., 1994) around the urban counties with the cubic root of their adjusted population base as weight was used as guidance to define the gravitational field. The cubic root is necessary to dampen the otherwise overall dominance of Berlin (East). See Figure 10.6 for multiplicative weighted Voronoi diagrams based on different powers of the population base. These diagrams have been calculated with GAM-BINI (Tiefelsdorf and Boots, 1997), which is a Windows utility program that can be integrated into Geographic Information Systems. The rural counties within the domain of an urban area were assigned to it. The administrative system of the former GDR was strictly hierarchically organized. This is also reflected by a permanent inflow of the workforce into Berlin (East). Therefore, the urban counties have direct ties to Berlin (East), which is the terminal knot in the hi-

---

[7] On the aggregation level of the 15 "Regierungsbezirke" including Berlin (East) the available migration matrix for 1987 indicates that 10 "Regierungsbezirke" have had strongest ties to Berlin (East), three have had second strongest and one had third strongest ties to Berlin (East). These four exceptions were all attached to "Regierungsbezirke" with direct links to Berlin (East). See Staatliche Zentralverwaltung für Statistik (1988, p 364) for data details and Holmes and Haggett (1977) for this graph theoretic mode of interaction table analysis.

Table 10.2. Distributional parameters of Moran's I under the assumption of spatial independence for an adjacency structure and a spatial hierarchy

	Spatial Structure	
	Adjacency	Hierarchy
observed I_0	0.1829	0.3531
$\Pr(I \leq I_0)$	> 0.9999	> 0.9999
number of links	1102	436
expectation	−0.0144	−0.0093
variance	0.0017	0.0041
skewness	0.1324	−0.0158
kurtosis	3.0225	3.2124
number of joint links	98	
correlation	0.1854	

erarchy. This approximated two level hierarchical migration system is shown in Figure 10.5. Note that it is not easy to spot by visual inspection potential spatial autocorrelation in hierarchical graphs like this one.

10.2.3   Testing for Inherent Spatial Autocorrelation

While this hypothesized migration system is formulated uni-directionally (i.e., from rural counties to urban areas to Berlin (East)) one can also expect, at a somehow lesser degree, return-migration along the same pathways. Therefore, the spatial migration link matrix has been parameterized bidirectionally. For the spatial hierarchy the distribution of links is highly asymmetric with only one link for each rural county and 50 links for Berlin (East). Thus, to control for this topology induced heterogeneity, the S-coding scheme needs to be employed. Table 10.2 gives details on both spatial structures and their associated Moran's I tests under the assumption of spatial independence.

For the bladder cancer model Moran's I indicates that for either an adjacency spatial link matrix or the hierarchical migration link matrix, the zero hypothesis has to be rejected. Consequently, there is a significant tendency in the regression residuals to cluster under either definition of the spatial link matrix which makes the ordinary least squares estimator inefficient. Furthermore, since both observed Moran's Is are highly significant and slightly correlated, it is impossible to distinguish under the zero hypothesis which process embraces the observed spatial dependence more accurately. The correlation between both Moran's Is is not surprising because their spatial structures share 98 links.[8] The spatial pattern in the regression residuals will be tested in Section 11.3 again by Moran's I conditional on the forces of either a spatial process defined on the adjacency structure or on the hierarchical migration structure.

---

[8] Note the asymmetry in the proportions. While 25% links in the hierarchical graph are joint links, only 10% of the links in the adjacency structures are shared with the hierarchical graph.

Chapter 11

A Spatial Analysis of the Cancer Data

This chapter aims at demonstrating the feasibility and flexibility of the exact conditional distribution of Moran's I as well as its usefulness in an applied framework. In detail, the spatial properties of the bladder cancer model will be analyzed. First the observed value I_0 of Moran's I needs to be linked to the most characteristic autocorrelation level ρ of an underlying spatial reference process. This link, which allows the identification of the most feasible spatial process, will be outlined in the next subsection. Having determined the underlying autocorrelation level of the spatial reference process, the distribution of both local Moran's I_i and global Moran's I conditional on the forces of the global spatial reference process can be given. In general terms, the conditional distribution of Moran's I measures the stochastic deviation of the observed Moran's I around a reference process, where Moran's I is defined on a specific spatial structure under question, the reference process is defined on an alternative spatial structure.

Using the conditional distribution of local Moran's I_i local heterogeneities in the residual structure of the bladder cancer model can be identified. For this analysis the underlying spatial process is defined on an adjacency relation of the counties of the former GDR. With the same approach, the probability of an observed value I_0 of global Moran's I defined on the adjacency structure conditional on a generating spatial process defined on the hierarchical migration pattern and vice versa can be given. This pair-wise comparison allows one to delineate the individual contribution of each spatial process and to identify the most likely spatial process generating the observed spatial pattern in the regression residuals of the bladder cancer model. Common to all employed analyses are the use of a binary spatial structure in the S-coding scheme and the assumption that the underlying spatial process is a Gaussian autoregressive spatial process.

11.1 Linking Moran's I to an Autoregressive Spatial Process

Figure 11.1 gives the exact conditional expectations of Moran's I subject to the autocorrelation levels ρ of an underlying hypothetical Gaussian autoregressive spatial process defined either on the adjacency or the hierarchical spatial structure. In both cases the conditional expectation of Moran's I increases monotonically with ρ. Note that for ρ close to the limits of its feasible range the conditional expectation of Moran's I will also be close to its bounds (see Table 11.1 for the exact bounds). Thus it can be shown that the conditional distribution of Moran's I close to either limit will be highly skewed away from either bound.

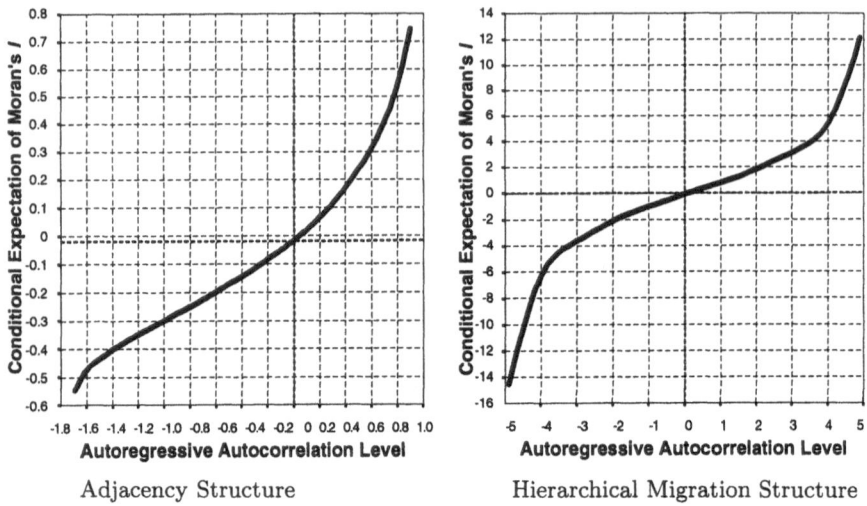

Adjacency Structure   Hierarchical Migration Structure

Fig. 11.1. Conditional expectations of Moran's I depending on an adjacency or a hierarchical spatial structure for an autoregressive spatial process.

Consequently it can not be approximated under these circumstances by the normal distribution.

Notice that binary spatial structures defined on a spatial hierarchy have a symmetric spectrum of eigenvalues with exactly $2 \cdot k$ nonzero eigenvalues where k is the number of separate branches in the hierarchy. In fact, each spatial structure related to an individual branch of the spatial hierarchy resembles the structure of a local link matrix.[1] That is, the graphs of spatial hierarchies and local link matrices are characterized by the absence of multiple paths between the spatial objects. Thus a spatial structure of a hierarchy is primarily aspatial and can not be used to unequivocally define the relative position of the spatial objects towards each other.

The graph of the conditional expectation of Moran's I provides the direct and so far unknown 'one-to-one match' (Anselin, 1995, p 108) between Moran's I and ρ. Under the premise that the observed value I_0 of Moran's I, which is a random variable, is representative for the underlying spatial process we can write

$$I_0 = \mathrm{E}[I \mid \mathbf{\Omega}(\rho_0)] \; . \qquad (11.1)$$

Other representative statistics measuring central tendencies like the median are also conceivable. While I_0 is known in empirical investigations we have to solve equation (11.1) for ρ_0. This can be done by a bisection search or any other search

---

[1] It is worthwhile to issue a note of caution. Most sparse spatial graphs like those reflecting higher order spatial lags (see Hepple, 1998), spatial hierarchies like the one employed in this monograph, and local spatial link matrices exhibit a kurtosis beyond 3 even under the assumption of spatial independence. These heavy tails in the distribution prohibit an approximation of the distribution by the four-parameter Pearson distribution family.

and interpolation technique. Table 11.1 gives for both spatial link matrices the related I_0 and ρ_0 values of an underlying autoregressive spatial process as well as their related feasible ranges. This linking procedure is in line with Sawa's (1978) application of the conditional moments to evaluate bias of the least squares estimator for the serial autocorrelation.

Table 11.1. Link between the observed values of Moran's I and the underlying autocorrelation parameters for an autoregressive spatial process in either an adjacency or a hierarchical spatial structure

Spatial Structure	one-to-one link		valid parameter ranges	
	observed I_0	matching ρ_0	I	ρ
Adjaceny:	0.1829	0.4197	$[-0.5894, 1.0467]$	$[-1.7053, 0.9116]$
Hierarchy:	0.3531	0.3194	$[-2.3093, 2.1042]$	$[-0.5007, 0.5007]$

11.2 Exploratory Analysis by means of Local Moran's I_i

Two extensive examinations on the properties and interpretation of local indicators of spatial association in recent papers by Ord and Getis (1995) and Anselin (1995) demonstrate their epistemological potential in exploratory and confirmatory spatial analysis of geo-referenced data. However, these authors raise several questions and outlined a research agenda with respect to (i) the so far unknown exact unconditional and conditional distribution of local indicators, (ii) the statistical consequences of the multiple test procedure[2] on the same data set by means of local indicators, (iii) the correlation between the local test statistics as well as (iv) the 'too large sample' problem[3] (see Kennedy, 1985, p 62). The attention in this section is focused on the calculation of the exact conditional probability of local Moran's I_i. Only if this distribution is available the other challenging subjects can be addressed. Furthermore, preliminary results on the correlation between the local Moran's I_i can be given and some properties of the test statistic with respect to extreme regression residuals can be discussed.

---

[2] Assuming the zero hypothesis of global independence is valid we still would expect a proportion α of local tests to exceed the critical value merely by chance. Thus the significance level α must be adjusted which is usually done by the Bonferroni correction (see Ord and Getis, 1995, p 298 for a discussion in the context of local spatial tests or Griffith and Amrhein, 1997, p 116, for a detailed statistical exposition). In essence, in the Bonferroni correction the overall desired significance level α is divided by the number of comparisons.

[3] In large samples it might well be the case under the assumption of spatial independence that we observe significant global spatial association whereas each local indicator does not deviate significantly from its expectation. This is very likely to happen for moderate global association especially when the Bonferroni correction for multiple tests is used.

The test statistic local Moran's I_i is a member of the class of local indicators of spatial association. These local statistics allow one to identify local variations in the overall pattern of spatial association and to probe the global stationarity assumption. These local variations in the data might be linked to unknown discrete spatial regimes of extreme data or "hot spots" around putative sources or spatial outliers. Further examples on the use of local indicators in Geography and Regional Science, for instance, can be found in O'Loughlin et al. (1994), who analyzed the regional and structural variations in the Reichstag election of the Weimar Republic in 1930, or Getis and Ord (1992), who studied sudden infant death syndrome in the counties in North Carolina as well as dwelling unit prices in metropolitan San Diego by zip-code districts, or Tiefelsdorf, Fortheringham, and Boots (conference on "Local Modelling with Spatial Data" in Newcastle-upon-Tyne, July 1997), who probed the local patterns in limiting-long-term-illness in 595 wards of north-east England. Furthermore, the journals "American Journal of Epidemiology" (1990) and "Statistics in Medicine" (1993, 1995 and 1996) devoted whole issues to the identification of disease clusters in particular and disease mapping in general. One of the major themes in these journals was the comparison of local statistics and mapping methods.

11.2.1   Local Moran's I_i as Ratio of Quadratic Forms

The key to derive firm statistical results on local Moran's I_i distribution is to define it as ratio of quadratic forms. So far, except for some initial work (Tiefelsdorf and Boots, 1997), the exact distribution of local Moran's I_i is unknown (Anselin, 1995, p 99). The definition which will be given here is algebraically equivalent with the one given in Anselin (1995, p 99) for local Moran's I_i as weighted sum by

$$I_i = s_i \cdot \frac{\widehat{\varepsilon}_i \cdot \sum_{j=1}^{n} b_{ij} \cdot \widehat{\varepsilon}_j}{\widehat{\varepsilon}^T \cdot \widehat{\varepsilon}} , \tag{11.2}$$

with s_i being an object specific scaling coefficient. For the S-coding scheme it is defined by

$$s_i = \frac{n^2}{2 \cdot \sqrt{d_i} \cdot \sum_{j=1}^{n} \sqrt{d_j}} \tag{11.3}$$

where d_i denotes the local linkage degree of the spatial object i. The suggested conditional randomization approach (Anselin, 1995, as well as Bao and Henry, 1996) to obtain local Moran's I_i's distribution is not helpful because it switches the status of a local observation. While an observation at the reference location i is supposed to be non-random, it becomes a random variable when attached as a satellite to any other reference location.

The local Moran's I_i statistic is defined as RQF by

$$I_i = \frac{\widehat{\varepsilon}^T \cdot (s_i \cdot \mathbf{B}_i) \cdot \widehat{\varepsilon}}{\widehat{\varepsilon}^T \cdot \widehat{\varepsilon}} \tag{11.4}$$

where the $n \times n$ symmetric *star-shaped* matrix $\mathbf{B}_i$ is specified by the i-th row and column of the global binary spatial link matrix $\mathbf{B}$ and with its remaining components set to zero, i.e.,

$$
\mathbf{B}_i = \begin{pmatrix}
0 & \cdots & 0 & b_{1i} & 0 & \cdots & 0 \\
\vdots & \ddots & \vdots & \vdots & \vdots & \ddots & \vdots \\
0 & \cdots & 0 & b_{i-1,i} & 0 & \cdots & 0 \\
b_{i1} & \cdots & b_{i,i-1} & 0 & b_{i,i+1} & \cdots & b_{in} \\
0 & \cdots & 0 & b_{i+1,i} & 0 & \cdots & 0 \\
\vdots & \ddots & \vdots & \vdots & \vdots & \ddots & \vdots \\
0 & \cdots & 0 & b_{ni} & 0 & \cdots & 0
\end{pmatrix} .
\tag{11.5}
$$

Details on the specification and properties of local Moran's I_i for the W-coding scheme and the C-coding scheme can be found in Tiefelsdorf and Boots (1997). For star-shaped spatial link matrices Cliff and Ord (1981, p 50) have shown that Moran's I cannot be normally distributed because it possesses an excessive kurtosis.

The overall set of n local Moran's I_is should satisfy Anselin's (1995) *additivity requirement* to identify spatial objects which are influential on the global statistics. It states that global Moran's I equals the average of the local Moran's I_is, i.e.,

$$
I = \frac{\sum_{i=1}^{n} I_i}{n} .
\tag{11.6}
$$

Thus there is a direct association between the local components I_i and the global Moran's I statistic. This enables us to identify spatial objects which are influential on the global autocorrelation statistic. Furthermore, this link between the local indicators and the global statistic allows one to investigate the exact distributions of the n local Moran's I_is conditional on a common global spatial process. Significant local Moran's I_is mark outstanding local heterogeneities in this global spatial process stemming from local non-stationarities. These may be based either on local clusters in regression residuals with equal signs leading to positive local autocorrelation or they are related to spatial outliers which do not match their surrounding environment. Note that the scaling coefficient s_i guarantees that the additivity requirement is satisfied for the S-coding (Tiefelsdorf, Griffith and Boots, 1999).

11.2.2   The Conditional Distribution of Local Moran's I_i

In this subsection the exact distribution of local Moran's I_i will be adjusted to the forces of a global spatial reference process. A treatment of this global impact has been considered important by both Anselin (1995) and Ord and Getis (1995) because neglecting its presence biases the search for locally outstanding clusters and outliers. The conditional distribution of local Moran's I_i measures the stochastic variation around the global spatial process and is therefore an indicator of local heterogeneities in the global spatial process. In contrast, the unconditional distribution of the local Moran's I_is just measures which spatial objects have a significant impact on the global spatial process. The conditional distribution of local Moran's I_i is easily obtained. First, the ρ_0 parameter associated with the observed global Moran's I_0 has to be determined by the method of the conditional expectation outlined in Section 11.1. Then, the conditional distribution local Moran's I_i can be given by $\Pr(I_i \leq I_{i0} \mid \mathbf{\Omega}(\rho_0))$ and calculated based on equation (7.9). Note, that the underlying spatial process $\mathbf{\Omega}$ is specified on the global link matrix $\mathbf{S}$ whereas the local test statistic I_i utilizes the spatial link matrix $s_i \cdot \mathbf{B}_i$.

Figure 11.2 displays for the bladder cancer example the spatial distribution of the exact probabilities of the local Moran's I_is conditional on an autoregressive spatial process defined on the adjacency matrix in the S-coding scheme with an autocorrelation level $\rho_0 = 0.4197$. To distinguish if positive local autocorrelation was induced by a spatial clustering in either negative or positive regression residuals, wedges mark the sign and magnitude of these residuals. Positive local autocorrelation (i.e., $\Pr(I_i \leq I_{i0} \mid \mathbf{\Omega}(\rho_0)) \geq 0.95$) in either positive or negative regression residuals clearly shows the tendency to cluster spatially. By the nature of the test statistic these clusters of positive autocorrelation in regression residuals of the same sign reach beyond the significant spatial object. Otherwise it would not be possible for local Moran's I_i at the reference location to indicate positive local spatial clustering. This indicates that for these counties and their neighborhoods some protective elements (for negative regression residuals) or risk factors (for positive regression residuals) may be present which operate on a larger spatial scale (i.e., clinical clusters). This can be seen, for instance, in the southwest corner of the former GDR with a spatial cluster of positive regression residuals and a spatial cluster of negative regression residuals or for Rostock in the north.[4] In contrast, negative local autocorrelation (i.e., $\Pr(I_i \leq I_{i0} \mid \mathbf{\Omega}(\rho_0)) \leq 0.05$) occurs mostly in isolation and pinpoints counties which are potential outliers with respect to their surrounding ensemble of counties. For these outstanding counties research must focus on the special nature of the observed local disease rates. For instance, unreliable data due to registration errors or highly localized risk factors (i.e., hot spots) or protect elements may cause such outliers. Interestingly, the urban counties are not disproportionately represented in the spatial outlier category. This indicates that the regression model accounts well for the urban/rural dichotomy in the cancer incidences.

The conditional probabilities of local Moran's I_i measure the local variations around the global spatial process. Therefore, as Figure 11.3 illustrates, the conditional probabilities are approximately uniformly distributed.[5] In contrast, if the global tendency of the regression residuals to cluster is ignored, the uncon-

---

[4] It is known that working in metal processing industries increases the risk for bladder cancer. In Rostock major shipyards are located which might partially explain the increased risk observed in and around Rostock.

[5] This also indicates that the autocorrelation level $\rho_0 = 0.4197$ represents the underlying spatial process very well.

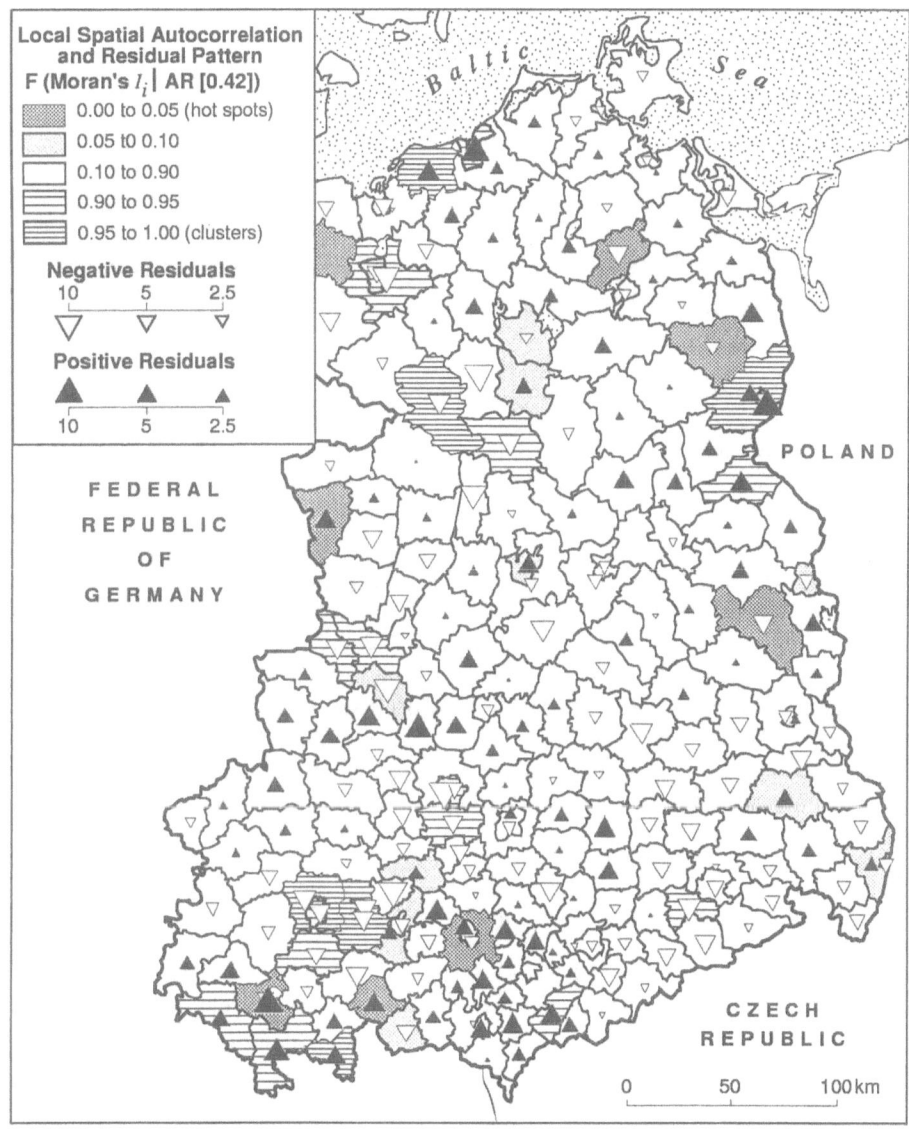

Fig. 11.2. Spatial distribution of the conditional probabilities of local Moran's I_i and the residuals from the male bladder cancer model

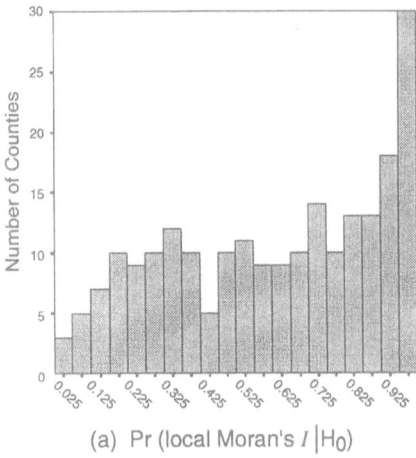
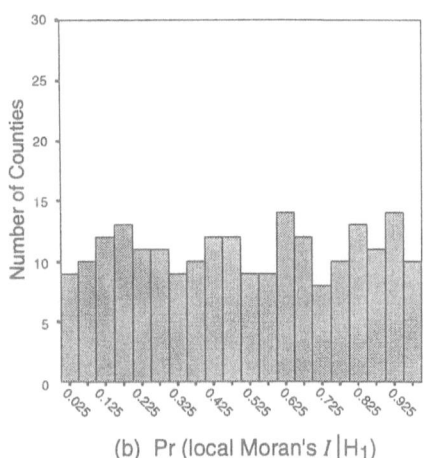

Fig. 11.3. Comparison of the distributions of $\Pr(I_i \leq I_{i0} \mid \sigma^2 \cdot \mathbf{I})$ and $\Pr(I_i \leq I_{i0} \mid \Omega(\rho_0 = 0.4197))$ for the 219 counties of the former GDR

ditional probabilities show an excess of significant positive local autocorrelation and only modest significant negative local autocorrelation and the search for outstanding spatial objects is biased. In particular, if we want to conduct a comparison of map patterns across several variables under investigation we need to use the conditional probabilities of local Moran's I_i because the single map pattern may be based on spatial processes exhibiting different global autocorrelation levels. For instance, in spatial epidemiology it is common practice to compare spatial patterns defined on the female and male sub-populations or on different accounting periods.

11.2.3   Correlation between Local Moran's I_i

The significance of the single local Moran's I_is can not be interpreted independently from each other because they are mutually correlated. This mutual correlation stems in part from the fact that the local Moran's I_is are defined on adjacent reference locations. That is, at spatial lag 1 they mutually share satellites as indicated in Figure 11.4. Even at spatial lag 2 an 'in-between' spatial object ties two local Moran's I_is together. Furthermore, as demonstrated by the Laurent series expansion of the autoregressive spatial process (see equation 4.21) spill-over effects can be expected even for higher order spatial lags. However, usually only the correlation between local Moran's I_is a few spatial lags apart needs to be considered because for a moderate autocorrelation level ρ the correlation approaches zero rapidly.

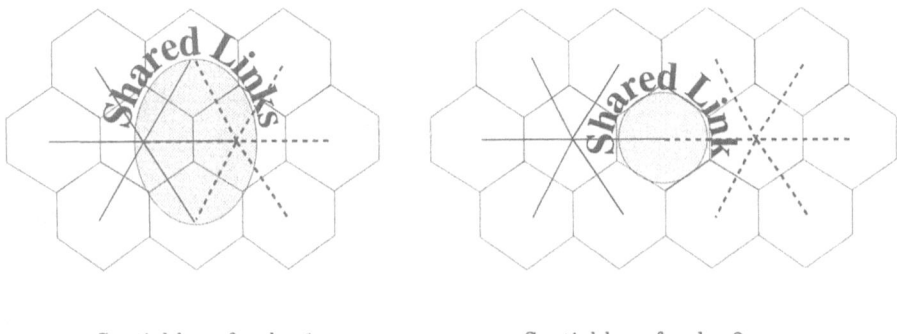

Spatial lag of order 1   Spatial lag of order 2

Fig. 11.4. Schematic topological influence on the correlation between local Moran's I_is depending on the spatial separation of their reference locations

Figure 11.5 illustrates the correlation of local Moran's I_is for the bladder cancer model under the assumptions of global spatial independence. The conditional correlation between local Moran's I_is can be evaluated with the method outlined in Chapter 9.3, however, this would induce heavy numerical calculations. The extreme high positive correlations at spatial lag 1 are attributable to the urban counties in the spatial tessellation of the former GDR. Most urban counties are embedded in a rural county and thus have only one link. Consequently, the observed value of local Moran's I_i for these urban counties depends largely on the value of local Moran's I_i for its surrounding rural county. Typically at spatial lag 2 we observe slight negative correlation between the local Moran's I_is. Negative correlation has also been observed by Ord and Getis (1995). Even at the spatial lag 4 there remains some minor spatial autocorrelation between the local Moran's I_is.

11.2.4   Impact of the Local Regression Residuals

Any statistical measure based on cross-products like the numerator in equation (11.2) is sensitive to extreme observations. The first component $\widehat{\varepsilon}_i$ is the regression residual at the reference location itself whereas the second component $\sum_{j=1}^{n} b_{ij} \cdot \widehat{\varepsilon}_j$ is a weighted sum around the reference location. First, since both components are multiplicatively linked, an extreme residual at the reference location or an extreme residual sum in its vicinity will have a strong impact on local Moran's I_i. Second, if either component in equation (11.2) is close to zero, local Moran's I_i can not be significant.

To illustrate the second property of local Moran's I_i Figure 11.6 gives for the GDR example a scatterplot of the regression residuals at the reference lo-

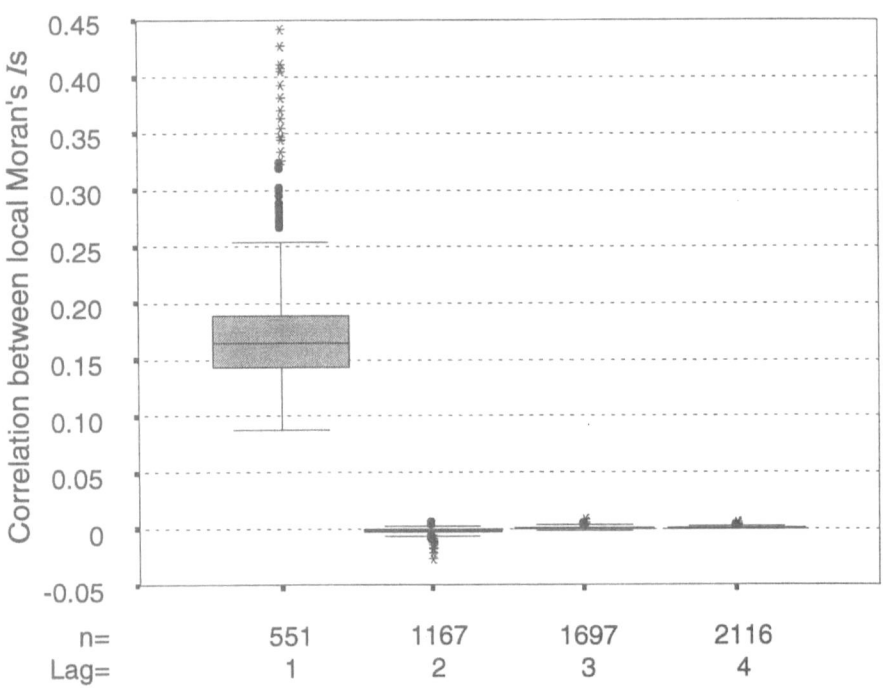

Fig. 11.5. Correlation between local Moran's I_is several spatial lags apart for the bladder cancer model under the assumption of spatial independence

cation against the exact conditional significances of the local Moran's I_is. Note the X-shaped pattern where all approximately zero regression residuals at the reference location induce non-significant local autocorrelation. Similarly, a plot of the residual sums around the reference location against the probability of local Moran's I_i also displays this X-shaped pattern. The exact significances have been chosen because, unlike the observed values of the I_i-statistic, they measure the extremity of local Moran's I_i independent of the coding scheme employed. Furthermore, the significances account for the effects of any local heterogeneity and independent variables. Therefore, they rest on a common scale and permit comparisons across the spatial objects.

The residual map and conditional significance in Figure 11.2 already presented the dominance of extreme regression residuals on local Moran's I_i. Confounding the impact of the residual at the reference location with those around the reference location is a major concern with local Moran's I_i. One way to overcome this problem is to strengthen the impact of the spatial objects around the reference location. Hence, we would have to forego the aspatial star-shaped spatial link matrix and apply a structure such as the wheel-shaped spatial relation

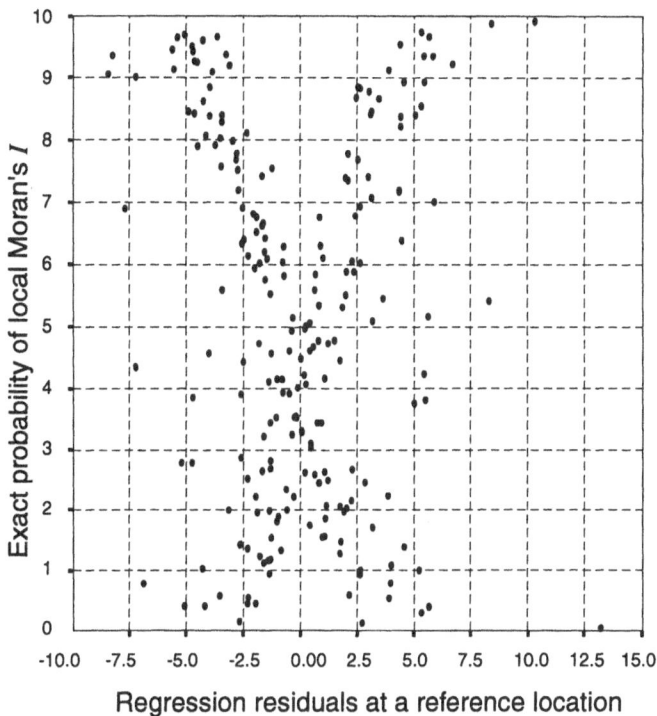

Fig. 11.6. The impact of the residual magnitudes at the reference location on the significance of local Moran's I_i for the bladder cancer model

graph (see Tinkler, 1972, p 33). In wheel-shaped spatial structures, the satellites are additionally connected to their neighbors and therefore their influence is strengthened relative to that of the reference location. While such a change would lead to a loss of the additivity property of local Moran's I_i, it would show a higher inter-correlation of adjacent local Moran's I_is and, accordingly, a smoother spatial pattern in the potential clusters. First experiments seem to be promising.

From my point of view local Moran's I_i defined on the sparse star-shaped spatial link matrix is most valuable to detect in an exploratory sense spatial outliers at the reference locations. In this respect local Moran's I_i helps well to satisfy Cressie's (1991, p 33) guideline that "analyzers of spatial data should be suspicious of observations when they are unusual with respect to their neighbors". Besides the identification of these local outliers, negative local spatial autocorrelation points also at areas where spatial interpolators based on the premise of global positive autocorrelation do not represent the data well.

11.3   The Identification of a Prevailing Spatial Process

As indicated already in Chapter 10.2 based on the low observed significance levels ($\alpha < 0.0001$) of global Moran's I specified either on an adjacency structure or on the hierarchical migration graph, it is impossible to distinguish which spatial process governs the data generating mechanism in the bladder cancer model. Consequently, we have either to assume that both schemes are operative and/or strongly correlated. Therefore, under these circumstances Haggett's (1976, see also Cliff and Ord, 1981, pp 24–27) exploratory decision rule, the more the observed data agree with a single hypothetical spatial process the less likely will the assumption of spatial independence be erroneously rejected, is not applicable. One way to address this identification problem is to use Bayesian techniques (see for instance, Hepple 1995a and 1995b). Alternatively, as applied here, the *power* function and the conditional perspective may help to isolate the individual impact of both spatial processes on the observed map pattern. The conditional perspective rests on a *non-nested* test strategy. It needs to be pointed out that an adjustment of the significance levels is again required due to multiple comparisons. This can be done, for instance, by Bonferroni bounds. Anselin (1986, and 1988, Chap. 13) gives a thorough discussion on the use, properties and problems associated with non-nested tests aimed at determining the specification of the spatial structure. However, in contrast to Anselin's discussion (1988, p 239), here non-symmetric pairwise test results are not a problem and, in fact, can be interpreted substantially, as well as the problems due to the asymptotic nature of his proposed tests do not occur because the presented method is exact.

Equation (4.10) models the linking mechanisms between a random input vector and the realized spatial pattern of a stochastic spatial process defined on a single spatial structure. These linking mechanisms can be extended. Brandsma and Ketellapper (1979) discuss a *biparametric* autoregressive specification of a spatial process in terms of two spatial structures with their associated autocorrelation levels ρ. If the underlying map pattern in the regression residuals of the bladder cancer model were generated by the adjacency structure $\mathbf{S}^{ADJA}$ as well as the hierarchical structure $\mathbf{S}^{HIER}$, then the joint linking mechanism is defined by the biparametric model

$$\varepsilon = (\mathbf{I} - \rho^{ADJA} \cdot \mathbf{S}^{ADJA} - \rho^{HIER} \cdot \mathbf{S}^{HIER})^{-1} \cdot \boldsymbol{\delta} . \qquad (11.7)$$

Note that in the black-box model of a spatial process in Figure 4.1 the question changes from 'which spatial process' to 'which share of each spatial process'. Clearly, if either ρ^{ADJA} or ρ^{HIER} equal zero model (11.7) reduces to a uniparametric model. While Brandsma and Ketellaper used a simultaneous likelihood ratio test, which is an asymptotic principle, to identify the composition of a spatial process, this section will introduce an exact small sample procedure based Moran's I conditional probabilities. Also Moran's I's exact power function is used to support the argument.

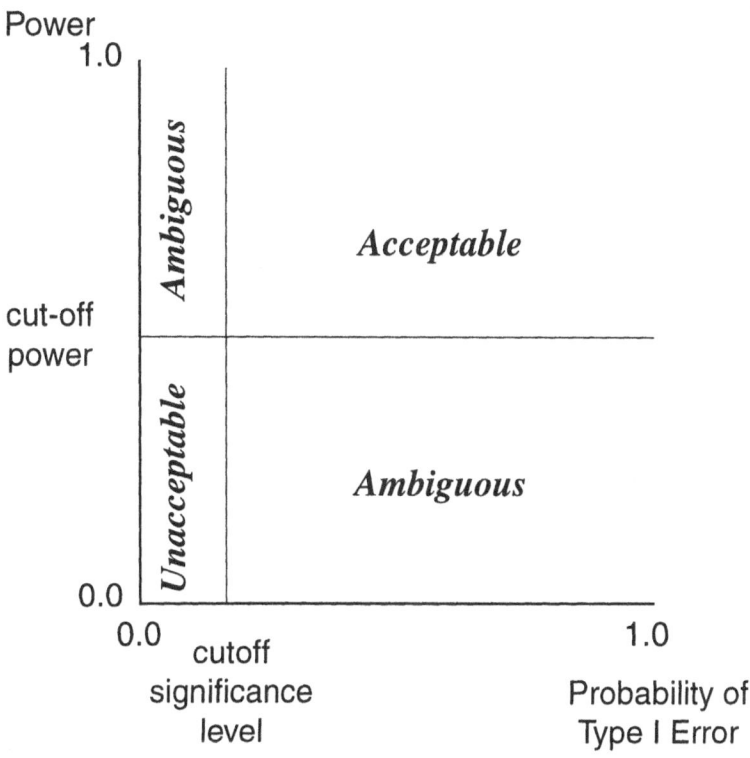

Fig. 11.7. Schematic relations between type I and type II error probabilities of a statistical test

11.3.1   The Power Function of Global Moran's *I*

Following Bollen's (1989, pp 347–349) argument the statistical power of a test can help the researcher to assess the compatibility of its outcome with either the underlying zero hypothesis or an alternative hypothesis. A low power of Moran's I defined on a specific spatial structure means that it is difficult for the autocorrelation test to reject the zero hypothesis. Therefore, a significant test result and a low power stress that the observed data are not compatible with the zero hypothesis (i.e., H_0 is unacceptable). In contrast, a high power but an insignificant test result suggest that under the perspective of the alternative hypothesis the data are consistent with the zero hypothesis (i.e., H_0 is acceptable). The remaining two combinations of high and low significances and powers are ambiguous. However, one should not conclude from this line of argument that tests with low power are preferable; it merely states that if a test with low power rejects the zero hypothesis then there really must be something substantial going on in the data. Figure 11.7 summarizes the relation between the significance and the power of a statistical test.

The Moran's I test statistics defined either on $\mathbf{S}^{ADJA}$ or $\mathbf{S}^{HIER}$ are highly significant in the bladder cancer model. Figure 11.8 gives the exact power functions of I^{ADJA} and I^{HIER} of a one-sided test against positive autocorrelation for an autoregressive spatial process in the S-coding scheme at a significance level $\alpha = 0.01$. The power function $\pi(\rho^{ADJA})$ for I^{ADJA} is defined by

$$\pi(\rho^{ADJA}) = 1 - Pr\big(I^{ADJA} \le I_\alpha^{ADJA} \mid \mathbf{\Omega}^{ADJA}(\rho^{ADJA})\big) \tag{11.8}$$

with the critical value $I_\alpha^{ADJA} = 0.0866$ and $\rho^{ADJA} \in [0.0000, 0.9116[$. The power function for the hierarchical spatial process is analogously defined with the critical value $I_\alpha^{HIER} = 0.1425$ and $\rho^{HIER} \in [0.0000, 0.5007[$. The abscissas in Figure 11.8 have been equally scaled by the ranges 0 to the upper bounds of ρ^{ADJA} or ρ^{HIER}, respectively. Within its feasible range the power function of I^{ADJA} is steeper than that of I^{HIER}. Thus I^{HIER} is less sensitive than I^{ADJA} to deviations from the zero hypothesis and a significant test result based on a hierarchical migration graph is germane.

11.3.2   Identification of a Spatial Process by a Non-Nested Test

While the power function allows the assessment of the *potential* of Moran's I to detect a spatial pattern, the conditional perspective of Moran's I is exact and resides on an *explicit model* of an underlying spatial process. To identify a spatial process the spatial structure associated with Moran's I is specified differently from the spatial structure underlying the model of the spatial process $\mathbf{\Omega}(\rho_0)$. This model sets the reference level around which Moran's I under question is re-assessed. If Moran's I in question is in variance with the underlying reference model of the spatial process, its conditional distribution will indicate a significant departure for the observed value I_0. Thus the underlying reference model of the spatial process is not able to capture all the inherent spatial association measured by Moran's I in question. Conversely, if the observed value I_0 of Moran's I is not significant with respect to its conditional distribution, its spatial structure will not introduce a substantial improvement into the reference model of the spatial process. However, due to the *non-symmetry* of the conditional perspective, the test should be repeated after switching the specification of the reference model with that of the Moran's I in question. Table 11.2 gives the four potential outcomes of these pairwise conditional tests for the situation that, under the zero hypothesis the individual Moran's Is are both significant and indistinguishable. If both conditional tests are insignificant either spatial structure virtually measures the same spatial association in the data and both can be used interchangeable. This happens if both processes are highly correlated due to the similarity of their associated spatial structures. Alternatively, if both conditional Moran's Is are significant, a combination of both spatial structures is operative and a biparametric specification of the process should be used.

To illustrate this switching procedure let us examine the probability of Moran's I^{ADJA} conditional on the spatial process $\mathbf{\Omega}^{HIER}$ and vice versa. The required parameters are given in Table 11.1. The conditional probability

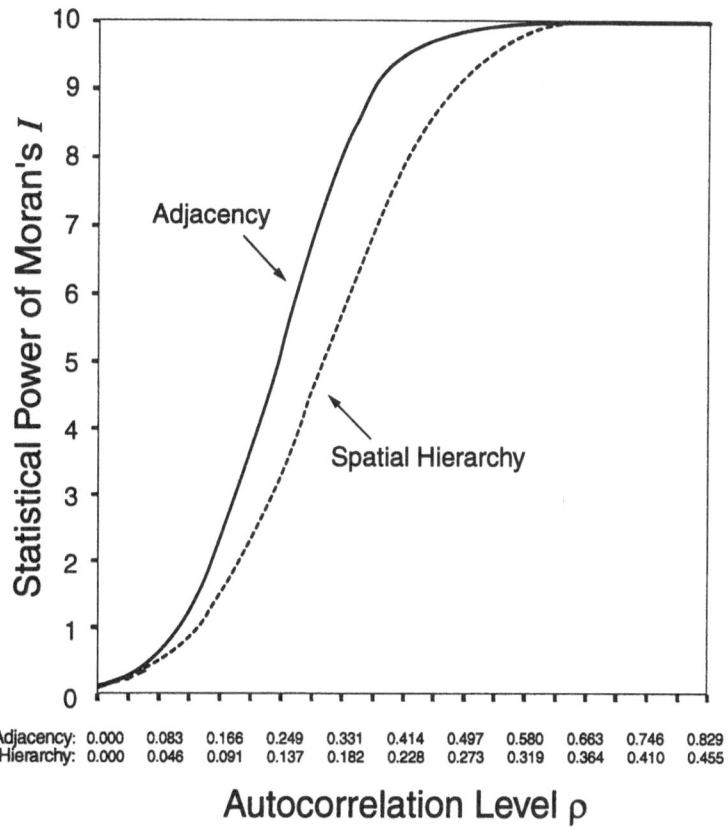

Fig. 11.8. The power functions of I^{ADJA} and I^{HIER} against positive spatial autocorrelation with a significance level $\alpha = 0.01$

$\Pr\left(I^{HIER} \leq 0.3531 \mid \Omega^{ADJA}(\rho_0^{ADJA} = 0.4197)\right) = 0.9994$ indicates that, given an underlying process defined on the adjacency structure, there is still a strong hierarchical association in the regression residuals. However, $\Pr\left(I^{ADJA} \leq 0.1829 \mid \Omega^{HIER}(\rho_0^{HIER} = 0.3194)\right) = 0.9549$ suggests that the hierarchical migration graph also captures most of the spatial association attributable to the spatial adjacency links. Thus, for this bladder cancer model we can conclude that the prevailing spatial association in the regression residuals has more in common with the hierarchical migration process than with neighborhood effects. Consequently, the disease incidence rate registered in one county is blurred by cases and risk factors originating from former places of residence.

Table 11.2. Decision rule for the non-nested comparison of two spatial processes under the conditional perspective

$F(I_0^{HIER} \mid \Omega^{ADJA})$	$F(I_0^{ADAJ} \mid \Omega^{HIER})$	
	insignificant	*significant*
insignificant	use either $\mathbf{S}^{ADAJ}$ or $\mathbf{S}^{HIER}$	use $\mathbf{S}^{ADAJ}$
significant	use $\mathbf{S}^{HIER}$	use the biparametric model specification

11.4 Discussion

This spatial analysis has revealed some hidden facts about male bladder cancer in the former GDR in particular and disease mapping in general. Ecological studies are able to identify risk factors, like tobacco abuse in the bladder cancer model. The use of local Moran's I_i allows one to identify potential spatial clusters where underlying risk factors may have been ignored and to pinpoint spatial outliers. As spatial outliers, the data may be questioned for accuracy and reliability. While the bladder cancer model seems to be not extremely influenced by the local non-stationarity it is still worth scrutinizing counties with significant local Moran's I_is. Due to the long latency period of cancers the observed local disease rates may be blurred by interregional migration. For male bladder cancer there is evidence that in fact the migration effect affected the data.

Some general comments on modeling disease rates in spatial epidemiology are helpful. In the initial data screening stage of non-rare diseases, models with a simple functional form are sufficient as long as the underlying model assumptions are satisfied. However, the model should already account for potential confounding factors and heterogeneities. Not only contagious diseases rest on spatial processes, it is also worthwhile to search for theoretically justified spatial processes in non-communicable diseases. In fact, an approach to handling the migration problem in spatial epidemiology has been proposed in this part of the monograph. The results based on a single population should be *consistent* with investigations of other populations (e.g., the opposite sex), different accounting periods or diseases sharing a comparable etiology. Only then is epidemiological significance gained on risk factors (Hennekens and Buring, 1987). In this respect, the map patterns of residuals from different analyses should be examined for consistency in their local clusters, local outliers and the underlying global spatial association.

With these considerations in mind a refined bladder cancer model should explicitly take into account autocorrelation based on a hierarchical migration graph. Eventually, the migration model should be broadened to incorporate channelized links and interurban connections (see Cadwallader, 1992). With additional explanatory variables the model must be scrutinized again for heteroscedasticity because the unexplained variation in the model errors decreases.

If the error variance turns out to depend on the population at risk then an autoregressive spatial version of Pocock, Cook, and Beresford's model (1981) could be calibrated. Clearly, such a model must be tackled by a maximum likelihood estimator. Finally, to assess, at least to some degree, the validity of the model, the analysis should be repeated for the female sub-population.

Several problems identified in the literature have been addressed and the guidelines for model validation issued by Anselin (1988, p 251) have been followed. The regression model has been tested thoroughly for the underlying assumptions of the OLS model. The employed small sample perspective did not leave the ambiguity of asymptotic results. The conditional distribution presented here allowed one to screen the global spatial relationships by means of local Moran's I_i for spatial heterogeneities and non-stationarities. The exact power function is available to assess the discriminating potential of a statistical test for spatial autocorrelation. A direct link between Moran's I and an underlying Gaussian process has been given by means of Moran's I conditional expectation. Potential spatial relationships have been explicitly stated and embedded in a theoretical framework. By means of the conditional distributions of the global Moran's Is the most feasible underlying spatial process was determined.

Chapter 12

Conclusions

Several major improvements to the current practice of identifying and analyzing spatial processes and their underlying spatial relationships have been laid out in this book:

The exact conditional distribution of Moran's I. Part II presented a closed theory on exact, small sample testing methods for spatial autocorrelation. In detail, the exact distribution function was derived under the forces of a global spatial process, its bounds and moments as well as cross-moments were given in analytical terms. This substantially reduces the need for simulation studies in Gaussian spatial processes. Part II's statistical content is a major unfolding of the results present in an initial paper by Tiefelsdorf and Boots (1995) on the [unconditional] exact distribution of Moran's I. This extensive discussion of statistical and technical matters, as well as the layout of ideas, were presented in this monograph with the intention of enabling spatial analysts to follow up the route of exact testing for spatial processes. More still needs to be done to improve the numerical efficiency in the calculation of the exact distribution by either using revised algorithms, for instance, by exploiting the sparseness of the spatial link matrix in the eigenvalue calculations (Pace and Barry, 1997), or by employing well-reasoned approximation techniques.

Exact test methods for spatial autocorrelation are under specific circumstances[1] mandatory and perform in a superior fashion. The paper by Hepple (1998) on exact tests for spatial autocorrelation and the editorial note by Thrift (1998) are further indications of the need of these exact methods.

The extremities of local Moran's I_i. The statistical results and the theoretical perspective of Part II enabled me to write local Moran's I_i as a ratio of quadratic forms in normal distributed regression residuals. Consequently, in contrast to simulation experiments, I was able to give its exact distribution, both under the assumption of spatial independence as well as conditional to a global spatial process. Knowledge of the exact distribution

---

[1] Recall: (i) the large kurtosis of local Moran's I_i's distribution even under the assumption of spatial independence, (ii) the non-normal distribution of Moran's I under alternative hypotheses, (iii) the exact power function of Moran's I, (iv) the identification of bi-parametric spatial processes, (v) the non-normal distribution for higher order spatial link matrices, (vi) the exact moments under alternative hypotheses with the identification of an underlying autocorrelation level and the pair-wise correlation between hypothetical spatial processes, as well as (vii) the marked non-normality even for empirical sample sizes around one hundred observations.

function allowed me to determine influential spatial regression residuals[2] on global Moran's I.

However, for local exploratory spatial data analysis, approximations of the exact distribution of local Moran's I_i will prove to be extremely helpful. The exact significance of local Moran's I_is need to be calculated for each switching reference location anew and their pair-wise correlation generates even more numerical efforts to be calculated at several spatial lags. Furthermore, an investigation of the correlation issue between the local Moran's I_is as well as the multiple testing strategy deserves further attention. Moreover, instead of concentrating in local autocorrelation analysis based on aspatial structures coded as sparse stars, one should explore the properties of local autocorrelation statistics related to wheel-shaped spatial structure.

This monograph goes far beyond the results presented in the note by Tiefelsdorf and Boots (1997) on the extremities of local Moran's I_i and their impact on global Moran's I under the zero hypothesis. However, that paper laid the foundation for the design of the variance stabilizing S-coding scheme. Knowledge of the exact moments of local Moran's I_i allowed us to perform further investigations on the local variance heterogeneities and to investigate the sensitivity of global Moran's I on the employed coding scheme.

A variance stabilizing coding scheme. With respect to the influence of the coding schemes on global Moran's I, one result of the article on the extremities of local Moran's I_i has been utilized in the paper on the variance stabilizing coding scheme for spatial link matrices by Tiefelsdorf, Griffith and Boots (1999). In fact, it allowed us to design analytically the variance stabilizing S-coding scheme which copes with the topologically induced heterogeneity. One can expect that the general design method as well as the particular S-coding scheme will be widely used in spatial analysis. For instance, with the presented method the modifiable areal unit problem can be re-addressed efficiently and studies can be conducted which are not biased by edge effects. I have the impression that it would be helpful to make other disciplines, like chemistry and physics, aware of the presented method. These disciplines also utilize link matrices in their research practice (see, for instance, Cvetković, Rowlingson and Simić, 1997).

This book uses the S-coding scheme in its empirical analysis to become independent from edge effects and the highly asymmetric distribution of the linkage degree in spatial hierarchy.

Practical applications of the conditional distribution. Part III gave some practical applications of the conditional distribution of Moran's I. Part III has demonstrated how to link by means of its conditional expectation an observed value of Moran's I to the autocorrelation level of a Gaussian

---

[2] In fact, it will be interesting to see in how far the results of a constrained linear programming approach applied by Boots and Dufournaud (1994) to identify critical links in a spatial links matrix which are associated with extreme regression residuals identified by local Moran's I_i.

spatial process. Use has been made of local Moran's I_i to identify hetero-geneities and outstanding spatial observations subject to a significant global spatial process. The exact power function of Moran's I was applied as a substantive indicator for an underlying spatial process. A pair-wise compar-ison procedure has been outlined to identify an underlying spatial process. Most importantly, however, a spatial structure has been defined by using substantial theoretical arguments. These arguments allowed me to link a hi-erarchical migration process with the spatial autocorrelation analysis of the cancer incidence rates.

While the statistical theory of testing for spatial relations and functional re-lationships in regression residuals by means of Moran's I is now well introduced in the geographical and statistical literature, it is my impression that a concise and closed theory of spatial structures is still lacking. Also, the link of spatial structures to spatial processes deserves further formal treatment. While Part I already cast a general epistemological framework on spatial structures and processes, much more needs to be done to raise the awareness of these intrin-sic spatial concepts. By focusing on spatial interrelationships – rather than on merely local characteristics or comparative studies – the discipline of geography as a whole will benefit. Spatial interrelations form a unique geographic perspec-tive which distinguishes Geography from other disciplines. Both *spatial auto-correlation* analysis and *spatial interaction* models are rooted in mutual spatial relationships between spatial objects. The empirical example in Part III of this monograph has demonstrated that spatially, autocorrelation analysis and spatial interaction modelling can supplement each other. In fact, in early geographical and statistical publications, the term "spatial interaction" was connoted with the statistical analysis of spatially autocorrelated processes (see, for example, Besag, 1974 and 1975, Ord, 1975, and Haining, 1978).

A valuable side effect of this process based perspective on spatial autocorrela-tion analysis is that the underlying spatial structure gains a substantial spatio-temporal meaning beyond conventional neighborhood arguments. Besides epi-demiological applications, the process based perspective can be applied also to address other spatial problems. For instance, Tiefelsdorf and Braun (1997) have used a sectorial spatial structure between urban districts to test for a misspecifi-cation of an intra-urban migration model. Theoretical results of urban geography state that the intra-urban migration pattern follows mostly a sectorial route from the center to the periphery and vice versa. If a migration model does not ac-count for this pattern, we will observe negative spatial autocorrelation between the sectors and positive spatial autocorrelation within the sectors. Visual inspec-tion of the adjusted residuals of a log-linear interaction model (see Tiefelsdorf and Boots, 1995) in fact indicated that Tiefelsdorf and Braun's interaction model was misspecified.

Hopefully, this process based perspective and other more substantial studies will pull spatial autocorrelation analysis out from its present marginal position within Geographical curriculum and academic practice. In part, the level of sta-tistical sophistication and lack of appropriate software can be blamed for this situation. However, it is my feeling that we are still without, with the exception of the initial attempt by Bailey and Gatrell (1995), good introductory textbooks to educate our students. Within the discipline of Geography, the aim should be, not to prove the mathematical and statistical sophistication of spatial autocorrelation analysis, but to demonstrate to Geographers and regional scientists the usefulness of such endeavors, to provide tools to conduct the analysis and to endow the basic understanding to interpret the results. In this respect, I hope that the empirical example of Part III serves these purposes. In fact, it has been demonstrated already that disease mapping and applications taken from epidemiology are rewarding geographical research areas ever since the seminal investigations initiated by Haggett in the Seventies.

Exact methods can readily be implemented into software packages as user-friendly black boxes and linked via macro languages and open interfaces directly to supporting desktop *Geographic Information Systems*.[3] The availability of exact methods reduces the burden for applied research which conducts simulations to assess the small sample properties of test results which might not even be asymptotically normal distributed (see Griffith and Lagona, 1998). In my experience, the exact methods presented here can be handled swiftly for moderate sized spatial data-sets (several hundred spatial objects) by using current high-end micro-computer equipment, while forthcoming generations of micro-computers will be able to 'crunch' larger data-sets comprising one thousand and more spatial objects with reasonable speed. Consequently, existing geo-referenced data sets, like those in cancer registries, are waiting to be mass-screened for spatial association and local heterogeneities.

The statistical theory presented in this book is rooted in the normal distribution and in regression residuals from simple ordinary least squares analysis. How far computationally attractive exact formulas for other distributions and alternative statistical models can be developed, is a subject for future research. Waller and Lawson (1995) have already derived the exact power function of a score test in Poisson distributed variables to detect focused clustering of diseases. The approach taken by Imhof (1961), who expressed the complex-valued inversion formula of a characteristic function in the real domain, is quite universal. However, the practicality of evaluating the joint-distribution function of two or more differently specified ratios of quadratic forms in spatial statistics still needs to be explored. The statistical theory, which is based on the bivariate or multivariate characteristic functions (see Stuart and Ord, 1994, pp 140ff), as well as first applications in the time-series domain, are already available (see Shephard, 1991, and Shively, Ansley and Kohn, 1990).

Two remarks stated more than 20 years ago by renowned statisticians in a special issue of the *The Statistician* on spatial data analysis still bear validity in current academic and geographical practice. Hammersley (1975) said that the

---

[3] In fact, Moran's I is implemented nowadays in most commercial Geographical Information Systems as well as in image processing systems to test for autocorrelation in spatial tessellations, by means of the normal approximation. Furthermore, Moran's I is the most frequently used test statistic to detect spatial autocorrelation.

"two-dimensional space is a wide open space for any statistician". I would add that the same is true for any Geographer focusing on spatial relations. Furthermore, Moran (1975) stated that "a great deal of statistical geography appears to me to be more descriptive than explanatory". To me, this statement only bears validity for non-relational geographical investigations. Finally, the comment by the statistician John Tukey (see Strauss, 1995) that "the best thing about being a statistician is that you get to play in everyone's back yard" is also easily transferable to Geography. For instance, while epidemiology and demography possess a rich methodology, both disciplines are largely lacking a spatial perspective. Geographers focusing on spatial relations can easily play an integral role in interdisciplinary research projects.

References

Abrahamse, A.P.J., and J. Koerts. A Comparison between the Power of the Durbin-Watson test and the Power of the BLUS test. *American Statistical Association Journal*, 64:938–948, 1969.

Agresti, A. *Categorical Data Analysis*. John Wiley, New York, 1990.

Ali, M.M. An Approximation to the Null Distribution and Power of the Durbin-Watson Statistic. *Biometrika*, 71:253–261, 1984.

Ali, M.M. Distributions of the Sample Autocorrelations when Observations are from a Stationary Autoregressive-Moving-Average Process. *Journal of Business, Economics, and Statistics*, 2:271–278, 1984.

Amrhein, C.G., and R. Flowerdew. The Effect of Data Aggregation on a Poisson Regression Model of Canadian Migration. *Environment and Planning A*, 24:1381–1391, 1992.

Anderson, T.W. On the Theory of Testing Serial Correlation. *Scandinavian Actuarial Journal*, 31:88–116, 1948.

Anselin, L. *Estimation Methods for Spatial Autoregressive Structures: A Study in Spatial Econometrics*. Regional Science Dissertation & Monograph Series #8, Ithaca, New York, 1980.

Anselin, L. A Note on Small Sample Properties of Estimators in a First-order Spatial Autoregressive Model. *Environment and Planning A*, 14:1023–1030, 1982.

Anselin, L. Non-nested Tests on the Weight Structure in Spatial Autoregressive Models: Some Monte Carlo Results. *Journal of Regional Science*, 26:267–284, 1986.

Anselin, L. Some Further Notes on Spatial Models and Regional Science. *Journal of Regional Science*, 26:799–802, 1986.

Anselin, L. Lagrange Multiplier Test Diagnostics for Spatial Dependence and Spatial Heterogeneity. *Geographical Analysis*, 20:1–17, 1988.

Anselin, L. *Spatial Econometrics: Methods and Models*. Kluwer Academic Publishers, Dordrecht, 1988.

Anselin, L. Local Indicators of Spatial Association – LISA. *Geographical Analysis*, 27:93–115, 1995.

Anselin, L., and D.A. Griffith. Do Spatial Effects Really Matter in Regression Analysis? *Papers of the Regional Science Association*, 65:11–34, 1988.

Anselin, L., and H.H. Kelejin. Testing for Spatial Error Autocorrelation in the Presence of Endogenous Regressors. *International Regional Science Review*, 1997.

Anselin, L., and O. Smirnov. Efficient Algorithms for Constructing Proper Higher Order Spatial Lag Operators. *Journal of Regional Science*, 36:67–89, 1996.

Anselin, L., and R. Florax. Small Sample Properties of Tests for Spatial Dependence in Regression Models: Some further Results. In Anselin, L., and R. Florax, editors, *New Directions in Spatial Econometrics*, pp 21–74. Springer Verlag, Berlin, 1995.

Anselin, L., and S. Rey. Properties of Tests for Spatial Dependence in Linear Regression Models. *Geographical Analysis*, 23:112–121, 1991.

Aptech Systems, I. *GAUSS – Version 3.2*. Aptech Systems, Inc., Maple Valley, WA, 1996.

Arbia, G. *Spatial Data Configuration in Statistical Analysis of Regional Economic and Related Problems*. Kluwer Academic Publishers, Dordrecht, 1989.

Bailey, T.C., and A.C. Gatrell. *Interactive Spatial Data Analysis*. Longman Group Ltd, Essex, 1995.

Bao, S., and M. Henry. Heterogeneity Issues in Local Measurements of Spatial Association. *Geographcial Systems*, 3:1–14, 1996.

Bartels, C.P.A., and L. Hordijk. On the Power of the Generalized Moran Contiguity Coefficient in Testing for Spatial Autocorrelation Among Regression Residuals. *Regional Science and Urban Economics*, 7:83–101, 1977.

Becker, N., and J. Wahrendorf. *Atlas of Cancer Mortality in the Federal Republic of Germany.* Springer Verlag, Berlin, 1997.

Besag, J.E. Spatial Interaction and the Statistical Analysis of Lattice Systems. *Journal of the Royal Statistical Society. Series B: Methodological*, 36:192–236, 1974.

Besag, J.E., and P.A.P. Moran. On the Estimation and Testing of Spatial Interaction in Gaussian Lattice Processes. *Biometrika*, 62:555–562, 1975.

Bivand, R.S. Regression Modeling with Spatial Dependence: An Application of Some Class Selection and Estimation Methods. *Geographical Analysis*, 16:25–37, 1984.

Bohman, H. A Method to Calculate the Distribution Function when the Characteristic Function is known. *Nordisk Tidskrift for Information Behandling*, 10:237–242, 1970.

Bohman, H. From Characteristic Function to Distribution Function via Fourier Analysis. *Nordisk Tidskrift for Information Behandling*, 12:279–283, 1972.

Bohman, H. Numerical Inversion of Characteristic Functions. *Scandinavian Actuarial Journal*, pp 121–124, 1975.

Bohman, H. Numerical Fourier Inversion. In Kahn, P.M., editor, *Computational Probability*. Academic Press, New York, 1980.

Bollen, K.A. *Structural Equations with Latent Variables.* John Wiley & Sons, New York, 1989.

Boots, B.N. Contact Number Properties in the Study of Cellular Networks. *Geographical Analysis*, 9:379–387, 1977.

Boots, B.N., and C. Dufournaud. A Programming Approach to Minimizing and Maximizing Spatial Autocorrelation Statistics. *Geographical Analysis*, 26:54–66, 1995.

Boots, B.N., and G.F. Rolye. A Conjecture on the Maximum Value of the Principal Eigenvalue of a Planar Graph. *Geographical Analysis*, 23:54–66, 1991.

Boots, B.N., and P. Kanaroglou. Incorporating the Effects of Spatial Structure in Discrete Choice Models of Migration. *Journal of Regional Science*, 28:495–507, 1988.

Borg, I., and J. Lingoes. *Multidimensional Similarity Structure Analysis.* Springer Verlag, Berlin, 1987.

Brandsma, A.S., and R.H. Ketellapper. A Biparametric Approach to Spatial Autocorrelation. *Environment and Planning A*, 11:51–58, 1979.

Brandsma, A.S., and R.H. Ketellapper. Further Evidence on Alternative Procedures for Testing of Spatial Autocorrelation Among Regression Disturbances. In Bartels, C.P.A., and R.H. Ketellapper, editors, *Exploratory and Explanatory Statistical Analysis of Spatial Data*, pp 113–136. Martinus Nijhoff Publishing, Boston, 1979.

Bundesministerium für Raumordnung, Bauwesen und Städtebau. *Raumordnungsbericht 1993.* Bundesministerium für Raumordnung, Bauwesen und Städtebau, Bonn, Germany, 1994.

Burridge, P. On the Cliff-Ord Test for Spatial Correlation. *Journal of the Royal Statistical Society. Series B: Methodological*, 42:107–108, 1980.

Burridge, P. Testing for a Common Factor in a Spatial Autoregression Model. *Environment and Planning A*, 13:795–800, 1981.

Buse, A. The Likelihood Ratio, Wald, and Lagrange Multiplier Tests: An Expository Note. *The American Statistican*, 36:153–157, 1982.

Cadwallader, M. *Migration and Residential Mobility. Macro and Micro Approaches.* The University of Wisconsin Press, Madison, 1992.

Caliper Inc. *Maptitude 4.0.* Newton, MA, 1997.

Cliff, A.D., and J.K. Ord. Testing for Spatial Autocorrelation Among Regression Residuals. *Geographical Analysis*, 4:267–284, 1972.

Cliff, A.D., and J.K. Ord. *Spatial Autocorrelation.* Pion, London, 1973.

Cliff, A.D., and J.K. Ord. The Choice of a Test for Spatial Autocorrelation. In Davis, J.C., and M.J. McCullagh, editors, *Display and Analysis of Spatial Data*, pp 54–77. John Wiley, Chichester, 1975.

Cliff, A.D., and J.K. Ord. *Spatial Processes. Models & Applications.* Pion Limited, London, 1981.

Cliff, A.D., P. Haggett, J.K. Ord, et al. *Elements of Spatial Structure. A Quantitative Approach.* Cambridge University Press, Cambridge, 1975.

Costanzo, C.M., L.J. Hubert, and R.G. Golledge. A Higher Moment for Spatial Statistics. *Geographical Analysis*, 15:347–351, 1983.

Cournant, R., and D. Hilbert. *Methods of Mathematical Physics.* Interscience Publishers, Inc., New York, 1953.

Cressie, N. *Statistics for Spatial Data.* John Wiley & Sons, Inc., New York, 1991.

Cvetkovic, D., P. Rowlinson, and S. Simic. *Eigenspaces of Graphs.* Cambridge University Press, Cambrigde, 1997.

Davis, B.M., and L.E. Borgman. Some Exact Sampling Distributions for Variogram Estimators. *Mathematical Geology*, 11:643–653, 1979.

de Gooijer, J.G. Exact Moments of the Sample Autocorrelations from Series Generated by General ARIMA processes of order (p, d, q), $d = 0$ or 1. *Journal of Econometrics*, 14:365–379, 1980.

de Jong, P., C. Sprenger, and F. van Veen. On Extreme Values of Moran's I and Geary's c. *Geographical Analysis*, 16:17–24, 1984.

Diggle, P.J., and P. Hall. A Fourier Approach to Nonparametric Deconvolution of a Density Estimate. *Journal of the Royal Statistical Society. Series B: Methodological*, 55:523–531, 1993.

Doll, R., C. Muir, and J. Waterhouse. *Cancer Incidence in Five Continents.* Springer Verlag, Berlin, 1970.

Doreian, P. On the Connectivity of Social Networks. *Journal of Mathematical Sociology*, 3:245–258, 1974.

Doreian, P., K. Teuter, and C.-H. Wang. Network Autocorrelation Models. *Sociological Methods and Research*, 13:155–200, 1984.

Dufour, J., and M.L. King. Optimal Invariant Tests for the Autocorrelation Coefficient in Linear Regression with Stationary or Nonstationary AR(1) errors. *Journal of Econometrics*, 47:115–143, 1991.

Durbin, J. Testing for Serial Correlation in Least Squares Regression when some of the Regressors are Lagged Dependent Variables. *Econometrica*, 38:410–421, 1970.

Durbin, J., and G.S. Watson. Testing for Serial Correlation in Least Squares Regression – I. *Biometrika*, 37:409–428, 1950.

Durbin, J., and G.S. Watson. Testing for Serial Correlation in Least Squares Regression – II. *Biometrika*, 38:159–178, 1951.

Durbin, J., and G.S. Watson. Testing for Serial Correlation in Least Squares Regression – III. *Biometrika*, 58:1–42, 1971.

Epps, T.W. Characteristic Functions and Their Empirical Counterparts: Geometrical Interpretations and Applications to Statistical Inference. *The American Statistican*, 47:33–38, 1993.

Evans, M.A., and M.L. King. Critical Value Approximations for Test of Linear Regression Disturbances. *Australian Journal of Statistics*, 27:68–83, 1985.

Farebrother, R.W. Pan's Procedure for the Tail Probabilities of the Durbin-Watson Statistics. Algorithm AS 153. *Journal of the Royal Statistical Society. Series C: Applied Statistics*, 29:224–227, 1980.

Farebrother, R.W. Corretion: Pan's Procedure for the Tail Probabilities of the Durbin-Watson Statistic. Algorithm AS 153. *Journal of the Royal Statistical Society. Series C: Applied Statistics*, 30:189, 1981.

Farebrother, R.W. The Distribution of a Linear Combination of Central χ^2 Random Variables. A Remark on AS 153: Pan's Procedure for the Tail Probabilities of the Durbin-Watson Statistic. *Journal of the Royal Statistical Society. Series C: Applied Statistics*, 33:366–369, 1984.

Farebrother, R.W. A Critique of Recent Methods for Computing the Distribution of the Durbin-Watson and other Invariant Test Statistics. *Statistische Hefte*, 35:365–369, 1994.

Ferguson, T.S. *Mathematical Statistics: A Decision Theoretic Approach*. Academic Press, New York, 1967.

Fomby, T.R., R.C. Hill, and S.R. Johnson. *Advanced Econometric Methods*. Springer Verlag, Berlin, 1984.

Gatrell, A.C. The Autocorrelation Structure of Central-place Populations in Southern Germany. In Wrigley, N., editor, *Statistical Applications in the Spatial Sciences*, pp 111–126. Pion Limited, London, 1979.

Gatrell, A.C. *Distance and Space. A Geographical Perspective*. Clarendon Press, Oxford, 1983.

Getis, A., and J.K. Ord. The Analysis of Spatial Association by Use of Distance Statistics. *Geographical Analysis*, 24:189–206, 1992.

Gil-Pelaez, J. A Note on the Inversion Theorem. *Biometrika*, 38:481–482, 1951.

Glick, B.J. The Spatial Organization of Cancer Mortality. *Annals of the Association of American Geographers*, 72:471–481, 1982.

Good, I.J. A Survey of the Use of the Fast Fourier Transform for Computing Distributions. *Journal of Statistical Computation and Simulation*, 28:87–93, 1987.

Goodchild, M.F. *Spatial Autocorrelation*. Geo Books, Norwick, 1986.

Greenberg, M. *Urbanization and Cancer Mortality: The United States Experience, 1950–1975*. Oxford University Press, New York, 1983.

Griffith, D.A. Towards a Theory of Spatial Statistics: Another Step Forward. *Geographical Analysis*, 19:69–82, 1987.

Griffith, D.A. *Advanced Spatial Statistics*. Kluwer, Dordrecht, 1988.

Griffith, D.A. What is Spatial Autocorrelation? Reflections on the past 25 Years of Spatial Statistics. *L'Espace Gographique*, 3:265–280, 1992.

Griffith, D.A. Spatial Autocorrelation and Eigenfunctions of the Geographic Weights Matrix Accompanying Geo-Referenced Data. *The Canadian Geographer*, 40:351–67, 1996.

Griffith, D.A., and C.G. Amrhein. *Multivariate Statistical Analysis for Geographers*. Prentice Hall, Upper Saddle River, 1997.

Griffith, D.A., and F. Lagona. On the Quality of Likelihood-Based Estimators in Spatial Autoregressive Models when the Data Dependence Structure is Misspecified. *Journal of Statistical Planning and Inference*, 69:153–174, 1998.

Haggett, P. Hybridizing Alternative Models of an Epidemic Diffusion Process. *Economic Geography*, 52:136–146, 1976.

Haining, R. Model Specification in Stationary Random Fields. *Geographical Analysis*, 9:107–129, 1977.

Haining, R. The Moving Average Model for Spatial Interaction. *Transactions and Papers, Institute of British Geographers, New Series*, 3:202–225, 1978.

Haining, R. Statistical Tests and Process Generators for Random Field Models. *Geographical Analysis*, 11:45–64, 1979.

Haining, R. Spatial Interdependencies in Population Distributions: A Study in Univariate Map Analysis. 2 Urban Population Distributions. *Environment and Planning A*, 13:85–96, 1981.

Haining, R. Spatial Models and Regional Science: A Comment on Anselin's Paper and Research Directions. *Journal of Regional Science*, 26:793–802, 1986.

Haining, R. *Spatial Data Analysis in the Social and Environmental Sciences*. Cambridge University Press, Cambridge, 1990.

Haining, R. Estimation with Heteroscedastic and Correlated Errors: A Spatial Analysis of Intra-urban Mortality Data. *Papers in Regional Science*, 70:223–241, 1991.

Haining, R. Diagnostics for Regression Modeling in Spatial Econometrics. *Journal of Regional Science*, 34:325–341, 1994.

Hammersley, J.M. The Wide Open Spaces. *The Statistican*, 24:159–160, 1975.

Hennekens, C.H., J.E. Buring, and S.L. Mayrent. *Epidemiology in Medicine*. Little, Brown Company, Boston, 1987.

Henshaw, R.C., Jr. Testing Single-equation Least Squares Regression Models for Autocorrelated Disturbances. *Econometrica*, 34:646–660, 1966.

Henshaw, R.C., Jr. Errata: Testing Single-equation Least Squares Regression Models for Autocorrelated Distrubances. *Econometrica*, 35:626, 1967.

Hepple, L.W. Bayesian Techniques in Spatial and Network Econometrics: 1. Model Comparison and Posterior Odds. *Environment and Planning A*, 27:447–469, 1995.

Hepple, L.W. Bayesian Techniques in Spatial and Network Econometrics: 2. Computational Methods and Algorithms. *Environment and Planning A*, 27:615–644, 1995.

Hepple, L.W. Exact Testing for Spatial Autocorrelation among Regression Residuals. *Environment and Planning A*, 30:85–108, 1998.

Hoeffding, W. The Large-sample Power of Tests Based on Permutations of Observations. *Annals of Mathematical Statistics*, 23:169–192, 1952.

Holmes, J.H., and P. Haggett. Graph Theory Interpretation of Flow Matrices: A Note on Maximization Procedures for Identifying Significant Links. *Geographical Analysis*, 9:388–399, 1977.

Hsiao, C. *Analysis of Panel Data*. Cambridge University Press, Cambridge, 1986.

Imhof, J.P. Computing the Distribution of Quadratic Forms in Normal Variables. *Biometrika*, 48:419–426, 1961.

Jammer, M. *Concepts of Space. The History of Theories of Space in Physics*. Dover Publications, Inc., New York, 1993.

Jones, M.C. On Moments of Ratios of Quadratic Forms in Normal Variables. *Statistics & Probability Letters*, 6:129–136, 1987.

Kelejin, H.H., and D.P. Robinson. Spatial Correlation: A Suggested Alternative to the Autoregressive Model. In Anselin, L., and R. Florax, editors, *New Directions in Spatial Econometrics*, pp 75–95. Springer Verlag, Berlin, 1995.

Kennedy, P. *A Guide to Econometrics*. MIT Press, Massachusetts, 1985.

King, M.L. Robust Tests for Spherical Symmetry and Their Application to Least Squares Regression. *The Annals of Statistics*, 8:1265–1271, 1980.

King, M.L. A Small Sample Property of the Cliff-Ord Test for Spatial Correlation. *Journal of the Royal Statistical Society. Series B: Methodological*, 43:263–264, 1981.

King, M.L. The Alternative Durbin-Watson Test. An Assessment of Durbin and Watson's Choice of Test Statistics. *Journal of Econometrics*, 17:51–66, 1981.

King, M.L. The Durbin-Watson Test for Serial Correlation: Bounds for Regressions with Trend and/or Seasonal Dummy Variables. *Econometrica*, 49:1571–1581, 1981.

King, M.L. A Point Optimal Test for Autoregressive Disturbances. *Journal of Econometrics*, 27:21–37, 1985.

King, M.L. A Point Optimal Test for Moving Average Regression Disturbances. *Econometric Theory*, 1:211–222, 1985.

King, M.L. Testing for Autocorrelation in Linear Regression Models: A Survey. In King, M.L., and D.E.A. Giles, editors, *Specification Analysis in the Linear Model*, pp 19–73. Routledge & Kegan Paul, London and New York, 1987.

King, M.L. Introduction to Durbin and Watson (1950, 1951) Testing for Serial Correlation in Least Squares Regression. I, II. In Kotz, S., and N.L. Johnson, editors, *Breakthroughs in Statistics. Volume II: Methodology and Distributions*, pp 229–236. Springer Verlag, Berlin, 1992.

King, M.L., and G.H. Hillier. Locally Best Invariant Tests of the Error Covariance Matrix of the Linear Regression Model. *Journal of the Royal Statistical Society. Series B: Methodological*, 47:98–102, 1985.

King, M.L., and M.A. Evans. The Durbin-Watson Test and Cross-Sectional Data. *Economics Letters*, 18:31–34, 1985.

King, M.L., and P.X. Wu. Small-disturbance Asymptotics and the Durbin-Watson and Related Tests in the Dynamic Regression Model. *Journal of Econometrics*, 47:145–152, 1991.

King, P.E. Problems of Spatial Analysis in Geographic Epidemiology. *Social Science and Medicine*, 13 D:249–252, 1979.

Koerts, J., and A.P.J. Abrahamse. On the Power of the BLUS Procedure. *American Statistical Association Journal*, 63:1227–1236, 1968.

Koerts, J., and A.P.J. Abrahamse. *On the Theory and Application of the General Linear Model*. Rotterdam University Press, Rotterdam, 1969.

Laurini, R., and D. Thompson. *Fundamentals of Spatial Information Systems*. Academic Press, London, 1992.

Lieberman, O. A Laplace Approximation to the Moments of a Ratio of Quadratic Forms. *Biometrika*, 81:681–690, 1994.

Lvy, P. *Calcul des Probabilits*. Gauthier-Villars, Paris, 1925.

Magnus, J.R. On Certain Moments Relating to Ratios of Quadratic Forms in Normal Variables: Further Results. *Sankhya*, B52:1–13, 1990.

Marshall, R.J. A Review of Methods for the Statistical Analysis of Spatial Patterns of Disease. *Journal of the Royal Statistical Society. Series A: General*, 154:421–441, 1991.

McCleary, R., and R.A. Hay. *Applied Time Series Analysis for the Social Sciences*. Sage Publications, Inc., Beverly Hills, California, 1980.

Meade, M., J. Florin, and W.M. Gesler. *Medical Geography*. The Guilford Press, New York, 1988.

Mehnert, W.H., M. Smans, C.S. Muir, et al. *Atlas of Cancer Incidence in the Former German Democratic Republic 1978–1982*. Oxford University Press, New York, 1992.

Molho, I. Spatial Autocorrelation in British Unemployment. *Journal of Regional Science*, 35:675–688, 1995.

Mood, A.M., F.A. Graybill, and D.C. Boes. *Introduction to the Theory of Statistics*. McGraw-Hill, Inc., New York, 1974.

Moran, P.A.P. The Use of Statistics in Geography. *The Statistican*, 24:161–162, 1975.

Nasca, P.C., M.C. Mahoney, and P.E. Wolfgang. Population Density and Cancer Incidence Differentials in New York State, 1978–82. *Cancer Causes Control*, 3:7–15, 1992.

Nystuen, J.D., and M.F. Dacey. A Graph Theoretical Interpretation of Nodal Regions. *Regional Science Association, Papers and Proceedings*, 7:29–42, 1961.

Ochi, M.K. *Applied Probability and Stochastic Processes. In Engineering and Physical Science.* John Wiley, New York, 1990.

Oden, N.L. Assessing the Significance of a Spatial Correlogram. *Geographical Analysis,* 16:1–16, 1984.

Odland, J. *Spatial Autocorrelation.* Sage Publications, Newbury Park, California, 1988.

Okabe, A., and Y. Sadahiro. An Illusion of Spatial Hierarchy: Spatial Hierarchy in a Random Configuration. *Environment and Planning A,* 28:1533–1552, 1997.

Okabe, A., B.N. Boots, and K. Sugihara. *Spatial Tessellations. Concepts and Applications of Voronoi Diagrams.* John Wiley & Sons, Chichester, 1992.

Okabe, A., B.N. Boots, and K. Sugihara. Nearest Neighbourhood Operations with Generalized Voronoi Diagrams: a Review. *International Journal of Geographical Information Systems,* 8:43–71, 1994.

O'Loughlin, J., C. Flint, and L. Anselin. The Geography of the Nazi Vote: Context, Confession, and Class in the Reichstag Election of 1930. *Annals of the Association of American Geographers,* 84:351–380, 1994.

Openshaw, S. Ecological Fallacies and the Analysis of Areal Census Data. *Environment and Planning A,* 16:17–31, 1984.

Ord, J.K. Estimation Methods for Models of Spatial Interaction. *Journal of the American Statistical Association,* 70:120–126, 1975.

Ord, J.K., and A. Getis. Local Spatial Autocorrelation Statistics: Distributional Issues and an Application. *Geographical Analysis,* 27:286–306, 1995.

Pace, K.R., and R. Barry. Quick Computation of Spatial Autoregressive Estimators. *Geographical Analysis,* 29:232–247, 1997.

Pan, J. Distribution of the Noncircular Serial Correlation Coefficients. *American Mathematical Society and Institut of Mathematical Statistics. Selected Translations in Probability and Statistics,* 7:281–291, 1968.

Pitman, E.J.G. The 'Closest Estimates' of Statistical Parameters. *Proceedings of the Cambridge Philosophical Society,* 33:212–222, 1937.

Pocock, S.J., D.G. Cook, and S.A.A. Beresford. Regression of Area Mortality Rates on Explanatory Variables: What Weighting is Appropriate. *Journal of the Royal Statistical Society. Series C: Applied Statistics,* 30:286–295, 1981.

Press, W.H., B.P. Flannery, S.A. Teukolsky, et al. *Numerical Recipes in Pascal.* Cambridge University Press, Cambridge, 1989.

Sawa, T. Finite-sample Properties of the k-Class Estimator. *Econometrica,* 40:653–680, 1972.

Sawa, T. The Exact Moments of the Least Squares Estimator for the Autoregressive Model. *Journal of Econometrics,* 8:159–172, 1978.

Schön, D., and B. Bellach. Relation between Breast Cancer Incidence and the Environment, Lifestyle and Socio-economic Factors – An Ecological Study. *International Journal of Environmental Health Research,* 5:293–302, 1995.

Schön, D., J. Bertz, and H. Hoffmeister. *Bevölkerungsbezogene Krebsregister in der Bundesrepublik Deutschland. Band 2.* MMV Medizin Verlag, München, 1989.

Schön, D., J. Bertz, and H. Hoffmeister. *Bevölkerungsbezogene Krebsregister in der Bundesrepublik Deutschland.* MMV Medizin Verlag, München, 1995.

Schwenk, A.J., and R.J. Wilson. On Eigenvalues of a Graph. In Beineke, L.W., and R.J. Wilson, editors, *Selected Topics in Graph Theory,* pp 307–336. Academic Press, London, 1978.

Sen, A. Large Sample-Size Distribution of Statistics Used In Testing for Spatial Correlation. *Geographical Analysis,* 9:175–184, 1976.

Shephard, N.G. From Characteristic Function to Distribution Function: A Simple Framework for the Theory. *Econometric Theory,* 7:519–529, 1991.

Shephard, N.G. Numerical Integration Rules for Multivariate Inversion. *Journal of Statistical Computation and Simulation*, 39:37–46, 1991.

Shively, T.S., C.F. Ansley, and R. Kohn. Fast Evaluation of the Distribution of the Durbin-Watson and Other Invariant Test Statistics in Time Series Regression. *Journal of the American Statistical Association*, 85:676–685, 1990.

Silvey, S.D. The Lagrangian Multiplier Test. *Annals of Mathematical Statistics*, 30:389–407, 1959.

Smith, M.D. On the Expectation of a Ratio of Quadratic Forms in Normal Variables. *Journal of Multivariate Analysis*, 31:244–257, 1989.

SPSS Inc. *SPSS for Windows. Release 6.1.3.* Chicago, 1995.

Staatliche Zentral Verwaltung für Statistik. *Statistisches Jahrbuch der DDR 1988.* Staatsverlag der DDR, Berlin, 1988.

Stetzer, F. Specifying Weights in Spatial Forecasting Models: The Results of Some Experiments. *Environment and Planning A*, 82:571–584, 1982.

Strauss, S. A Statistical Man for all Seasons. *The Globe and Mail.* D8, Toronto, April 1, 1995.

Stuart, A., and J.K. Ord. *Kendall's Advanced Theory of Statistics. Distribution Theory.* Edward Arnold, London, 1994.

TCI Software Research. *Scientific Workplace 2.5.* Pacific Grove, CA, 1996.

Terui, N., and M. Kikuchi. The Size-adjusted Critical Region of Moran's I Test Statistics for Spatial Autocorrelation and Its Application to Geographical Areas. *Geographical Analysis*, 26:213–227, 1994.

Theil, H. The Analysis of Disturbances in Regression Analysis. *Journal of the American Statistical Association*, 60:1067–1079, 1965.

Thompson, S.K. *Sampling.* John Wiley & Sons, Inc., New York, 1992.

Thrift, N.J. Editorial Note. *Environment and Planning A*, 30:109, 1998.

Tiefelsdorf, M. Some Practical Applications of Moran's I's Exact Distribution. *Papers in Regional Science*, 77:101–129, 1998.

Tiefelsdorf, M., and B.N. Boots. The Specification of Constrained Interaction Models Using SPSS's Loglinear Procedure. *Geographical Systems*, 2:21–38, 1995.

Tiefelsdorf, M., and B.N. Boots. The Exact Distribution of Moran's I. *Environment and Planning A*, 27:985–999, 1995.

Tiefelsdorf, M., and B.N. Boots. Letters to the Editor. *Environment and Planning A*, 28:1900, 1996.

Tiefelsdorf, M., and B.N. Boots. A Note on the Extremities of Local Moran's I_i and their Impact on Global Moran's I. *Geographical Analysis*, 29:248–257, 1997.

Tiefelsdorf, M., and B.N. Boots. GAMBINI – A GIS Utility Programme to Calculate Multiplicatively Weighted Voronoi Diagrams. *http://www.wlu.ca/ ˜wwwgeog/gambini.htm*, 1997.

Tiefelsdorf, M., and G. Braun. The Migratory System of Berlin after Unification in the Context of Global Restructuring. *Geographica Polonica*, 69:23–44, 1997.

Tiefelsdorf, M., D.A. Griffith, and B.N. Boots. A Variance Stabilizing Coding Scheme for Spatial Link Matrices. *Environment and Planning A*, 31:165–180, 1998.

Tillman, J.A. The Power of the Durbin-Watson Test. *Econometrica*, 43:959–974, 1975.

Tinkler, K.J. The Physical Interpretation of Eigenfunctions of Dichotomous Matrices. *Transactions, Institute of British Geographers*, 55:17–46, 1972.

Tinkler, K.J. *An Introduction to Graph Theoretical Methods in Geography.* Geo Abstracts, Norwich, 1977.

Tobler, W. A Computer Movie Simulating Urban Growth in the Detroit Region. *Economic Geography, Supplement*, 46:234–240, 1970.

Upton, G. Information from Regional Data. In Griffith, D.A., editor, *Spatial Statistics: Past, Present, and Future*, pp 315–356. Department of Geography, Syracuse University, Syracuse, 1990.

Upton, G., and B. Fingleton. *Spatial Data Analysis by Example. Volume I: Point Pattern and Quantitative Data*. John Wiley, New York, 1985.

Upton, G., and B. Fingleton. *Spatial Data Analysis by Example. Volume II: Categorical and Directional Data*. John Wiley, New York, 1989.

Wackernagel, H. *Multivariate Geostatistics*. Springer Verlag, Berlin, 1995.

Waller, L.A. Does The Characteristic Function Numerically Distinguish Distributions. *The American Statistican*, 49:150–152, 1995.

Waller, L.A., and A.B. Lawson. The Power of Focused Tests to Detect Disease Clustering. *Statistics in Medicine*, 14:2291–2308, 1995.

Waller, L.A., B.W. Turnbull, and J.M. Hardin. On Obtaining Distribution Functions by Numerical Inversion of Characteristic Functions with Applications. *The American Statistican*, 49:346–350, 1995.

Walter, S.D. Assessing Spatial Pattern in Disease Rates. *Statistics in Medicine*, 12:1885–1894, 1993.

Wartenberg, D., and M. Greenberg. Solving the Cluster Puzzle: Clues to Follow and Pitfalls to Avoid. *Statistics in Medicine*, 12:1763–1770, 1993.

Wendt, H. Wanderungen nach und innerhalb von Deutschland unter besonderer Berücksichtigung der Ost-West Wanderungen. *Zeitschrift für Bevölkerungswissenschaft*, 19:517–540, 1994.

Young, G., and A.S. Householder. Discussion of a Set of Points in Terms of their Mutual Distances. *Psychometrika*, 3:19–22, 1938.

Index

Lecture Notes in Earth Sciences

For information about Vols. 1–19
please contact your bookseller or Springer-Verlag